Application of Manganese Pincer Complexes for Hydrogenation and Dehydrogenation Reactions

Von der Fakultät für Mathematik, Informatik und Naturwissenschaften der RWTH Aachen University zur Erlangung des akademischen Grades einer Doktorin der Naturwissenschaften genehmigte Dissertation

vorgelegt von

Viktoriia Zubar, M. Sc.

aus

Znob-Novgorodske

Berichter: *Univ.-Prof. Dr. rer. nat. Magnus Rueping*

            Prof. Dr. Frédéric Patureau

Tag der mündlichen Prüfung: *15. September 2020*

Viktoriia Zubar
Application of Manganese Pincer Complexes for Hydrogenation and Dehydrogenation Reactions

ISBN: 978-3-95886-374-3
1. Auflage 2020

Bibliografische Information der Deutschen Bibliothek
Die Deutsche Bibliothek verzeichnet diese Publikation in der Deutschen Nationalbibliografie; detaillierte bibliografische Daten sind im Internet über www.dnb.ddb.de abrufbar.

Das Werk einschließlich seiner Teile ist urheberrechtlich geschützt. Jede Verwendung ist ohne die Zustimmung des Herausgebers außerhalb der engen Grenzen des Urhebergesetzes unzulässig und strafbar. Das gilt insbesondere für Vervielfältigungen, Übersetzungen, Mikroverfilmungen und die Einspeicherung und Verarbeitung in elektronischen Systemen.

Vertrieb:

    © Verlagshaus Mainz GmbH Aachen
    Süsterfeldstr. 83, 52072 Aachen
    Tel. 0241 / 87 34 34 00
    www.Verlag-Mainz.de

Herstellung:

    Druckerei Mainz GmbH Aachen
    Süsterfeldstraße 83
    52072 Aachen
    www.DruckereiMainz.de

    Satz: nach Druckvorlage des Autors
    Umschlaggestaltung: Stephan Dammrau

    printed in Germany

    D82 (Diss. RWTH Aachen University, 2020)

Eidesstattliche Erklärung

Ich, Viktoriia Zubar, erkläre hiermit, dass diese Dissertation und die darin dargelegten Inhalte die eigenen sind und selbstständig, als Ergebnis der eigenen originären Forschung, generiert wurden.

Hiermit erkläre ich an Eides statt

1. Diese Arbeit wurde vollständig oder größtenteils in der Phase als Doktorand dieser Fakultät und Universität angefertigt;
2. Sofern irgendein Bestandteil dieser Dissertation zuvor für einen akademischen Abschluss oder eine andere Qualifikation an dieser oder einer anderen Institution verwendet wurde, wurde dies klar angezeigt;
3. Wenn immer andere eigene- oder Veröffentlichungen Dritter herangezogen wurden, wurden diese klar benannt;
4. Wenn aus anderen eigenen- oder Veröffentlichungen Dritter zitiert wurde, wurde stets die Quelle hierfür angegeben. Diese Dissertation ist vollständig meine eigene Arbeit, mit der Ausnahme solcher Zitate;
5. Alle wesentlichen Quellen von Unterstützung wurden benannt;
6. Wenn immer ein Teil dieser Dissertation auf der Zusammenarbeit mit anderen basiert, wurde von mir klar gekennzeichnet, was von anderen und was von mir selbst erarbeitet wurde;
7. Ein Teil oder Teile dieser Arbeit wurden zuvor veröffentlicht.

Aachen, 22.09.2020

(Ort, Datum)   (Unterschrift)

The work discussed in that thesis was conducted at the Institute of Organic Chemistry at RWTH Aachen under the supervision of Professor Magnus Rueping from January 2017 to February 2020.

Parts of this work have already been published:

V. Zubar, Y. Lebedev, L. M. Azofra, L. Cavallo, O. El-Sepelgy, M. Rueping, Hydrogenation of CO_2-Derived Carbonates and Polycarbonates to Methanol and Diols by Metal–Ligand Cooperative Manganese Catalysis, *Angew. Chem. Int. Ed.*, **2018**, *57*, 13439–13443.

V. Zubar, J. C. Borghs and M. Rueping, Hydrogenation or Dehydrogenation of *N*-Containing Heterocycles Catalysed by a Single Manganese Complex, *Org. Lett.* **2020**, *22* (10), 3974-3978.

L. M. Azofra, M. A. Tran, **V. Zubar**, L. Cavallo, M. Rueping and O. El-Sepelgy, Racemic alcohols to optically pure amine precursors enabled by catalyst dynamic kinetic resolution: experiment and computation, *Chem. Commun.*, **2020**, 56, 9094-9097.

V. Zubar, J. Sklyaruk, A. Brzozowska, M. Rueping, Chemoselective Hydrogenation of Alkynes to (*Z*)-Alkenes Using an Air-Stable Base Metal Catalyst, *Org. Lett.* **2020**, *22* (14) 5423–5428.

Unpublished manuscripts, which are based on this PhD-work:

V. Zubar, A. Dewanji, M. Rueping Chemoselective Hydrogenation of Nitroarenes using an Air Stable Manganese Catalyst, *submitted*

V. Zubar, A. Brzozowska and M. Rueping Intramolecular Aniline Alkylation for the Synthesis of diverse *N*-Heterocycles, *in preparation*

Other manuscripts:

J. Sklyaruk, **V. Zubar**, J. C. Borghs, M. Rueping, Methanol as the Hydrogen Source in the Selective Transfer Hydrogenation of Alkynes Enabled by a Manganese Pincer Complex, *Org. Lett.* **2020**, 22, 15, 6067–6071

J. C. Borghs, **V. Zubar**, L.M. Azofra, J. Sklyaruk, M. Rueping, Manganese-Catalyzed Regioselective Dehydrogenative C- versus N-Alkylation Enabled by a Solvent Switch: Experiment and Computation, *Org. Lett.* **2020**, *22*, 4222-4227

A. Brzozowska, **V. Zubar**, R.-C. Ganardi, M. Rueping, Chemoselective Hydroboration of Propargylic Alcohols and Amines Using a Manganese(II) Catalyst, *Org. Lett.* **2020**, *22* (10), 3765-3769.

M. Szewczyk, M. Magre, **V. Zubar**, M. Rueping, Reduction of Cyclic and Linear Organic Carbonates Using a Readily Available Magnesium Catalyst, *ACS Catal.* **2019**, *9*, 12, 11634-11639.

A. Brzozowska, L. M. Azofra, **V. Zubar**, I. Atodiresei, L. Cavallo, M. Rueping, O. El-Sepelgy, Highly Chemo- and Stereoselective Transfer Semihydrogenation of Alkynes Catalyzed by a Stable, Well-Defined Manganese(II) Complex, *ACS Catal.*, **2018**, *8* (5), 4103–4109.

O. El-Sepelgy, A. Brzozowska, J. Sklyaruk, Y. K. Jang, **V. Zubar**, M. Rueping, Cooperative Metal–Ligand Catalyzed Intramolecular Hydroamination and Hydroalkoxylation of Allenes Using a Stable Iron Catalyst, *Org. Lett.* **2018** *20* (3), 696-699.

Acknowledgment

I would like to thank my supervisor, Prof. Magnus Rueping, for the possibility to conduct my doctorate studies in his research group and for his support and valuable suggestions. I would also like to thank Prof. Patureau for being my second reviewer.

I would like to express my gratitude to my co-workers, in particular to Dr. Aleksandra Brzozowska, Dr. Yoon-Kyung Jang, Dr. Jan Sklyaruk, Dr. Osama El-Sepelgy, Dr. Luis Miguel Azofra, Dr. Jannik Borghs, Mai Anh Tran, Dr. Yury Lebedev and Prof. Luigi Cavallo for successful collaborations and productive discussions.

Besides, I greatly appreciate the assistance of Dr. Erli Sugiono, Dr. Iuliana Atodiresei, Cornelia Vermeeren and Gabriele Bertrand for the organisation of group activities, administrative work, proof-reading of publications and much else. Additionally, I want to thank Dr. Wolfgang Bettray and the entire analytical department for performing MS, IR and NMR measurements.

I would like to thank the entire Rueping group for creating a friendly atmosphere, especially Dr. Patricia Krach, Dr. Aleksandra Brzozowska, Dr. Abhishek Dewanji, Dr. Kevin Kaut, Dr. Alban Falconnet, Dr. Marcin Szewczyk, Dr. Marc Magre, Dr. Yoon-Kyung Jang, Dr. Watchara Srimontree, Tengfei Ji and Dr. Eva Paffenholz. Additionally, I would like to thank my lab-mates Dr. Jiaqi Jia and Julian Krischel for a great working experience. Many thanks to Dr. Aleksandra Brzozowska and Dr. Pavlo Nikolaienko for the great company in KAUST. Moreover, I highly appreciate the help of Dr. Aleksandra Brzozowska, Dr. Abhishek Dewanji and Dr. Alban Falconnet for corrections and suggestions to this thesis.

Furthermore, I want to thank my family, Stephan and his family for their unconditional support and encouragement during my studies.

To my husband and family

Table of Contents

Chapter 1: Introduction 1

1.1 Introduction to transition metal catalysis 1

1.2 Metal-ligand cooperative catalysis 3

1.3 Application of manganese complexes in homogeneous catalysis 5

    1.3.1 Manganese-catalysed oxidation reactions 5

    1.3.2 C-H activation catalysed by manganese complexes 5

    1.3.3 Application of manganese complexes for hydrogenation-dehydrogenation reactions ... 6

Chapter 2: Results and discussion 15

2.1 Manganese-catalysed hydrogenation of cyclic organic carbonates 15

    2.1.1 Introduction 15

    2.1.2 Optimization of reaction conditions 16

    2.1.3 Scope of substrates 18

    2.1.4 Mechanistic investigations 19

    2.1.6 Proposed reaction mechanism 20

    2.1.7 Summary 23

2.2 Diastereoselective amination of racemic alcohols 24

    2.2.1 Introduction 24

    2.2.2 Optimization of reaction conditions 25

    2.2.3 Scope of substrates 26

    2.2.4 Mechanistic studies 29

    2.2.5 Proposed reaction mechanism 30

    2.2.6 Summary 31

2.3 Hydrogenation of alkynes and alkenes catalysed by manganese pincer complexes 32

    2.3.1 Introduction 32

    2.3.2 Optimisation of reaction conditions 33

    2.3.3 Scope of substrates 34

2.3.4 Mechanistic studies 36

2.3.5 Proposed reaction mechanism 38

2.3.6 Summary 39

2.4 Hydrogenation and dehydrogenation of heterocycles with the application of manganese pincer complexes 40

    2.4.1 Introduction 40

    2.4.2 Optimisation of reaction conditions 41

    2.4.3 Scope of substrates 42

    2.4.4 Mechanistic studies and proposed reaction mechanism 44

    2.4.5 Summary 45

2.5 Manganese-catalysed hydrogenation of nitroarenes 46

    2.5.1 Introduction 46

    2.5.2 Optimisation of reaction conditions 47

    2.5.3 Scope of substrates 48

    2.5.4 Mechanistic studies 50

    2.5.5 Summary 52

2.6 Intermolecular alkylation of amines with alcohols to form heterocycles 53

    2.6.1 Introduction 53

    2.6.2 Optimisation of reaction conditions 54

    2.6.3 Scope of substrates 55

    2.6.4 Reaction mechanism 57

    2.6.5 Summary 58

2.7 Manganese-catalysed alkylation of nitroarenes with alcohols 59

    2.7.1 Introduction 59

    2.7.2 Optimisation of reaction conditions 59

    2.7.3 Scope of substrates 60

    2.7.4 Summary 61

Chapter 3: Summary and Outlook 62

Chapter 4: Experimental part 65

4.1 Manganese-catalysed hydrogenation of cyclic organic carbonates 65

    4.1.1 General information 65

    4.1.2 Ligand and complex synthesis and characterization 65

    4.1.3 General procedures and reaction analysis 71

4.2 Diastereoselective amination of racemic alcohols 77

    4.2.1 General information 77

    4.2.2 General procedures and reaction analysis 78

4.3 Hydrogenation of alkynes and alkenes catalysed by manganese pincer complex 82

    4.3.1 General information 82

    4.3.2 Ligand and complex synthesis and characterization 83

    4.3.3 General procedures and reaction analysis 85

4.4 Hydrogenation and dehydrogenation of heterocycles with the application of manganese pincer complexes 90

    4.4.1 General information 90

    4.4.2 General procedure and reaction analysis 91

4.5 Manganese-catalysed hydrogenation of nitroarenes 96

    4.5.1 General information 96

    4.5.2 Experimental procedure and characterizations of the products 97

4.6 Intermolecular alkylation of amines with alcohols to form heterocycles 101

    4.6.1 General information 101

    4.6.2 Experimental procedure and characterizations of the products 102

4.7 Manganese-catalysed alkylation of nitroarenes with alcohols 105

    4.7.1 General information 105

    4.7.2 Experimental procedure and characterizations of the products 106

Chapter 5: List of abbreviations 108

Chapter 6: References 110

Chapter 1: Introduction

1.1 Introduction to transition metal catalysis

On the way towards sustainable and green chemistry, scientists found a key technology which allows chemical transformations to proceed with minimum waste generation and energy consumption. The key is transition metal catalysis which is also attractive from an economical point of view. At the end of 19th century, Alfred Werner defined basics of coordination chemistry[1] which were significant for the development of transition metal catalysis. In the 1930s, the Oxo Process, in which olefins react with syngas to produce linear and branched aldehydes, was developed by Otto Roelen.[2] The reaction is still widely used and allows to produce millions metric tons of oxo chemicals. A following milestone in transition metal catalysis occurred in the 1950s when the Wacker Process was developed. The reaction allows to oxidize olefins to aldehydes and ketones with the use of $PdCl_2$ as catalyst.[3] Many other reactions were developed involving transition metal catalysis, including important processes such as C-C bond formations, site selective C-H bond activation, arylation of amines and alcohols and others which have been also applied in industrial processes to obtain bulk chemicals, fine chemicals and polymers. Another indicator highlighting the importance of homogenous transition metal catalysis is the acknowledgement by the Noble Prize committee. Thus, in 2001, the Nobel Prize in chemistry was awarded jointly to William S. Knowles, Ryoji Noyori and K. Barry Sharpless for their work on transition metal catalysed asymmetric hydrogenation and oxidation reactions. Later in 2005, Yves 0Chauvin, Robert H. Grubbs and Richard R. Schrock were selected for the Noble Prize in chemistry for their work on the development of the metathesis reaction in organic synthesis. And finally, recent Noble Prize in chemistry in 2010 was awarded to Richard F. Heck, Ei-ichi Negishi and Akira Suzuki for palladium-catalysed cross coupling reactions in organic synthesis. Until recently, reports featuring homogeneous transition metal catalysis were focused on using rare and expensive metals. The application of noble metals raises major concern, as they are toxic and consequently they produce toxic waste. And to their low availability (listed as critical raw materials for European Union)[4], noble metals require difficult extraction from the Earth's crust.

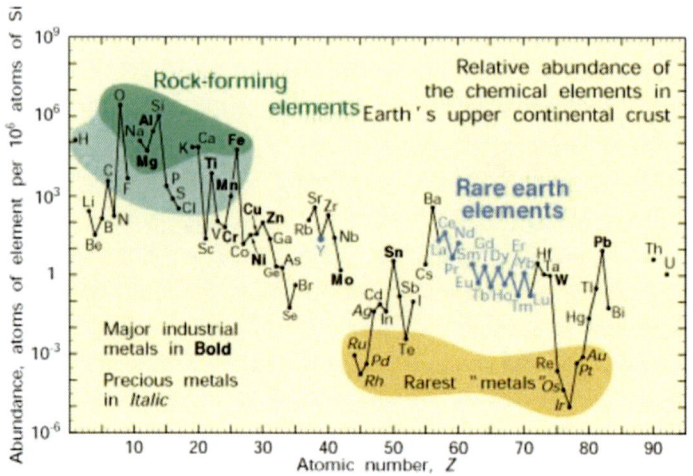

Figure 1. Abundance (atom fraction) of the chemical elements in Earth's crust as a function of atomic number. Source: https://pubs.usgs.gov/fs/2002/fs087-02/

Therefore, the replacement of noble-metal catalysts by earth-abundant, cheap alternatives nowadays is a topic of great interest. Although, the catalytic activity of noble metals in homogeneous (de-)hydrogenation reactions is very high, its toxicity for living organisms, price and low abundance led chemists to look for better alternatives, such as first row transition metals. The interest of chemical community shifted towards the development of new, sustainable and green catalysts based on Fe, Co, Ni and Mn which are significantly more abundant and less toxic compared to noble metals. For example, Fe and Mn belong to the group of metals with minimum safety concern regarding contaminations for pharmaceuticals.[5] Research in recent decades brought attention to the fact that base metal catalysts can sometimes outperform noble metal catalysts and even provide unexpected new products.

In this thesis, close attention was paid to manganese, as a representative of non-noble transition metals applied in homogeneous catalysis. Manganese is the 12th most abundant element and 5th most abundant metal in the Earth's crust. It is an essential component of over 30 enzymes for animals, and it also takes part in oxygen evolution as part of photosynthesis. Cost, abundance and relative low toxicity make manganese an attractive alternative to noble metals.

1.2 Metal-ligand cooperative catalysis

Transition metal catalysis is based on activation of the substrates by coordinating to the metal, which lowers the activation energy of the reaction. One of the major advantages of homogeneous transition metal catalysis compared to heterogeneous catalysis is easily tunable selectivity of the reaction by changing the metal or the electronic or steric properties of the ligand. The main difference of metal-ligand cooperative catalysis from classical transition metal catalysis is the participation of the ligand in the catalytic reaction. The concept of metal-ligand cooperation came from observing enzymes, particularly of the activation of hydrogen by hydrogenases.[6] This reaction proceeds *via* heterolytic cleavage of the H-H bond across the metal-ligand bond (Scheme 1). This approach inspired the establishment of metal-ligand cooperative catalysis and allowed to develop new catalytic transformations.

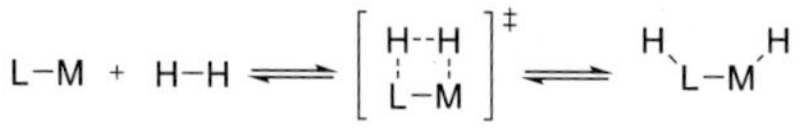

$$L-M + H-H \rightleftharpoons \left[\begin{array}{c} H--H \\ | \quad | \\ L-M \end{array} \right]^{\pm} \rightleftharpoons \begin{array}{c} H \quad \quad H \\ \diagdown \quad \diagup \\ L-M \end{array}$$

Scheme 1. Hydrogen activation by metal-ligand cooperation.

The metal-ligand cooperative catalysts are mainly distinguished by two types of complexes. One of the classic examples of the first type (M-L mode) is a ruthenium catalyst developed in 1995 by Noyori and co-workers.[7] The catalyst consists of two fragments, Ru–phosphine complex and an ethylenediamine ligand. It is one of the first examples where it is crucial to have an acidic NH group adjacent to the metal center which acts as proton shuttle.

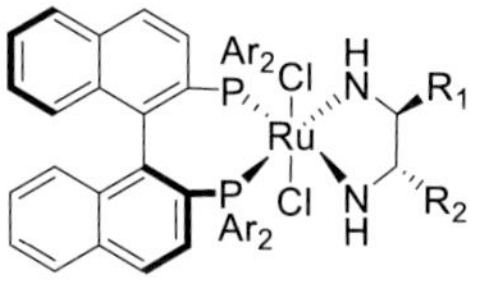

Figure 2. General structure of Ru catalyst applied for asymmetric hydrogenation and transfer hydrogenation of ketones developed by Noyori's group.

Many other catalysts could be classified into this group. Representative examples could be seen in the work of Morris.[8] Additionally to this group belong aliphatic tridentate pincer complexes. Pincer ligands are three-coordinate chelating agents that binds to neighbouring coplanar sites in a meridional configuration. The tridentate coordination of pincer ligand affords strong binding to the metal center and results in high thermal stability to the resulting complexes.[9]

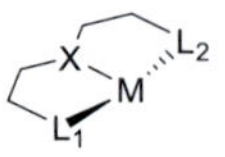

M: metal atom
X: central atom (N, C, Si)
L$_1$, L$_2$: side arms (PR$_2$, NR$_2$, SR, CR$_3$)

Figure 3. General structure of pincer ligands.

The aliphatic tridentate pincer ligands with a central amide/amine nitrogen donor are widespread in the field of metal-ligand cooperative catalysis. In the beginning of 1980s, Fryzuk et al. reported a reversible heterolytic splitting of H$_2$ utilizing iridium disilylamido PNP complexes for the first time.[10] Since then the field of aliphatic pincer complexes gained significant attention.[11] They proved to be active for hydrogenation of ketones, [12–14] aldehydes,[14] nitriles,[15] imines,[12] CO$_2$ for formate,[16] esters,[17,18] carbonates[19] and N-heterocycles.[20] -as well as dehydrogenation of alcohols,[13,18,21] formic acid,[22] heterocycles[20] and amine-borane adducts.[23] It was suggested by experimental and computational investigations that metal-ligand cooperation through a metal–amide/metal–amine bond may play an important role in these reactions.

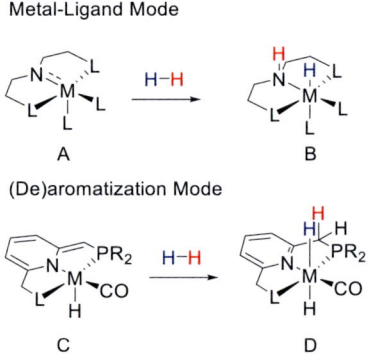

Scheme 2. Two major types of metal-ligand cooperation for H$_2$ activation.

Another type of metal-ligand cooperation involves activation through aromatization/dearomatization. To this group belong pincer complexes mostly based on the lutidine or 2-picoline fragment, which possesses at least one CH$_2$ group in *ortho* position. Upon the reaction with strong base, the pyridine component undergoes dearomatization creating exocyclic double bond which later can accept a proton and the metal centre can accept another unit of the reacting molecule. One of the major contributors to the topic is the group of David Milstein. A range of different catalysts were developed featuring this mode of activation providing catalytic activity for numerous reactions including hydrogenation and dehydrogenation reactions, C-H bond activation and dehydrogenative coupling reactions.[24] Many other researchers started investigating this topic. Thus, Morris and co-workers successfully applied the concept for the water splitting reaction.[25] Sanford and co-workers used lutidine based ruthenium complexes for the activation of CO_2 and carbonyl compounds.[26] Later on, the group of Pidko applied ruthenium catalysts for the hydrogenation of esters and CO_2.[27]

1.3 Application of manganese complexes in homogeneous catalysis

Manganese catalysts found a wide range of application in homogeneous catalysis. Among them, manganese carbonyl complexes with bipyridines as ligand were well studied for electrochemical CO_2 reduction.[28] Additionally, manganese is known to catalyse cross coupling reactions,[29] oxidation reactions and C-H activation as well as hydrogenation and dehydrogenation reactions.

1.3.1 Manganese-catalysed oxidation reactions

Until recently manganese complexes were mostly known and readily applied for the epoxidation of olefins (Scheme 3). Pioneering work of J. K. Kochi and co-workers[30] described racemic epoxidation of unfunctionalized olefins using a cationic Mn(III)-salen complex. Next, E. N. Jacobsen[31] and T. Katsuki[32] independently described enantioselective epoxidation using chiral Mn(III)-salen complexes.

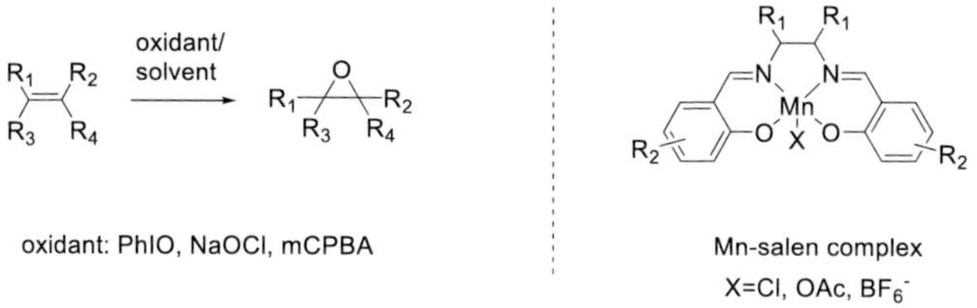

oxidant: PhIO, NaOCl, mCPBA

Mn-salen complex
X=Cl, OAc, BF_6^-

Scheme 3. Epoxidation of unfunctionalized olefins catalysed by Mn(III)-salen complex.

1.3.2 C-H activation catalysed by manganese complexes

Another significant tool in organic synthesis is manganese-catalysed C-H activation. Recently interest in this topic dramatically increased since it provides step-economical access to a range of important compounds for medicinal chemistry, agrochemical industry and other fields. To mimic the behaviour of manganese taking part in enzymatic processes, such as metabolism of carbohydrates, cholesterol, and amino acids in the human body, many catalytic systems using $Mn(CO)_5Br$ and $Mn_2(CO)_{10}$ as metal precursors were developed. The reaction usually proceeds through chelation assistance of a substrate.

The development in the field started with the work of Bruce and Stone[33] where they reported formation of manganacycle by stochiometric reaction of azobenzene with $MeMn(CO)_5$ *via* C-H activation (Scheme 4).

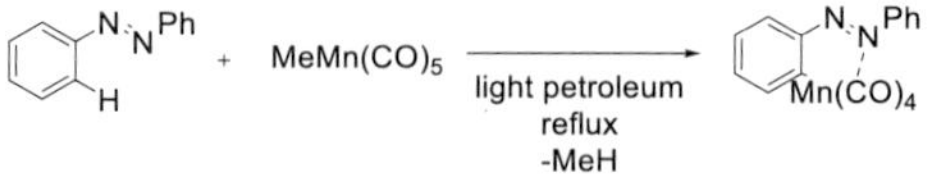

Scheme 4. Stochiometric C-H activation.

Furthermore, the stochiometric reaction inspired the development of catalytic C-H activation. Representative examples are shown in Scheme 5.[34] Our research group were also interested in the topic, and a few projects featured manganese-catalysed C-H activations. Among them selective C2-alkylation of indoles can be highlighted as a representative example of C-H functionalisation.[35]

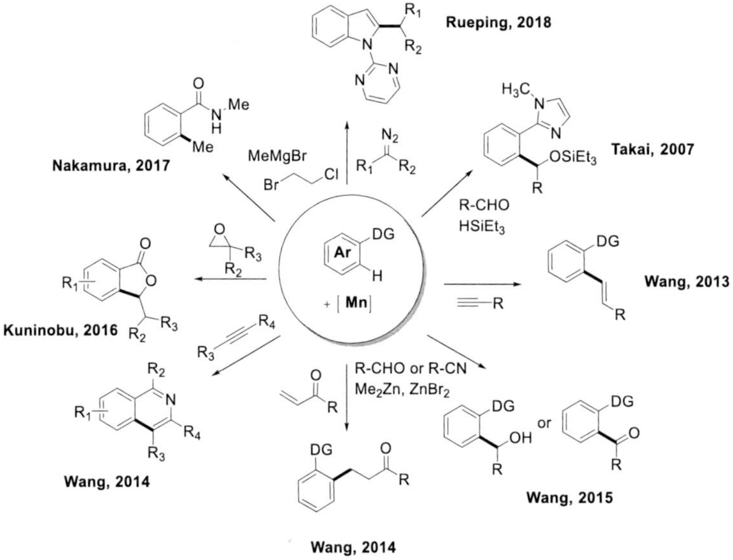

Scheme 5. Chosen examples of Mn-catalysed C-H activation.

1.3.3 Application of manganese complexes for hydrogenation-dehydrogenation reactions

Hydrogenation

Catalytic hydrogenation using molecular hydrogen is very important methodology in synthetic organic chemistry used in academia and industry. One of the major advantages of hydrogen as a reducing agent is its low price and high availability. Additionally, catalytic hydrogenation has very high atom economy and does not produce waste by-products. Non catalytic hydrogenation can occur only at very harsh reaction conditions, which is also true for heterogeneous catalysts. When homogeneous catalysts perform at relatively mild reaction conditions it usually allows higher selectivity and functional group tolerance. Thus, significant attention was drawn to the development of complexes for homogeneous catalysis which could be successfully applied for hydrogenation of organic molecules.[36] Most reports on homogeneous hydrogenation reactions have focused on noble metals. Due to low availability, price and toxicity of these metals, the development of base-metal systems is highly desired and currently gaining considerable interest.[37] The widespread development of manganese-catalysed hydrogenation reactions started in 2016. Pioneering works reported the hydrogenation of ketones, aldehydes, nitriles and esters. In 2016, Beller and co-workers[38] as well as Kempe and co-workers[39] independently presented results on hydrogenation of carbonyl compounds using well-defined Mn-PNP complexes. Later on, two publications from Sortais and co-workers[40] featured the application of Mn-PNP and Mn-PN complexes for the hydrogenation of ketones and aldehydes (Scheme 6).

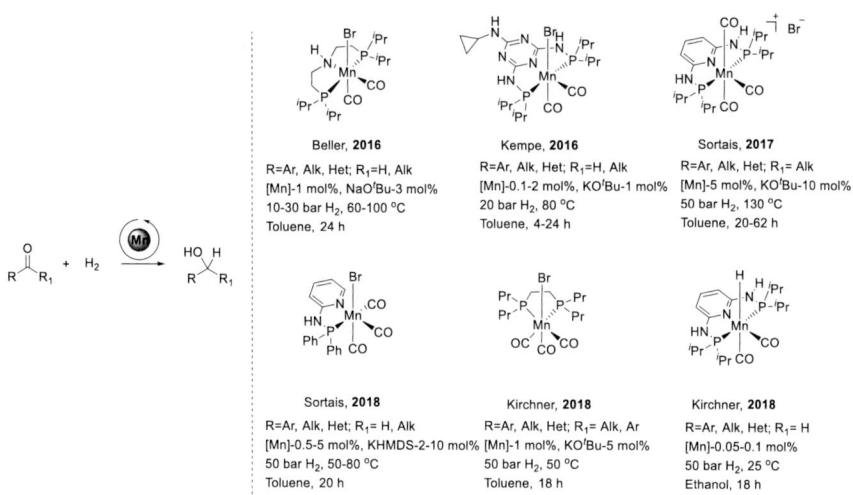

Scheme 6. Hydrogenation of aldehydes and ketones.

Next, Kirchner and co-workers applied biphosphine Mn(I) complex for a selective reduction of ketones[41] and activated Mn-PNP complex for the selective hydrogenation of aldehydes.[42] Meanwhile, the asymmetric hydrogenation of ketones using chiral manganese pincer complexes was developed.[43] Some of the above-mentioned catalytic systems were also active towards hydrogenation of nitriles. Beller and co-workers applied a Mn-^{iPr}PNP complex for this transformation.

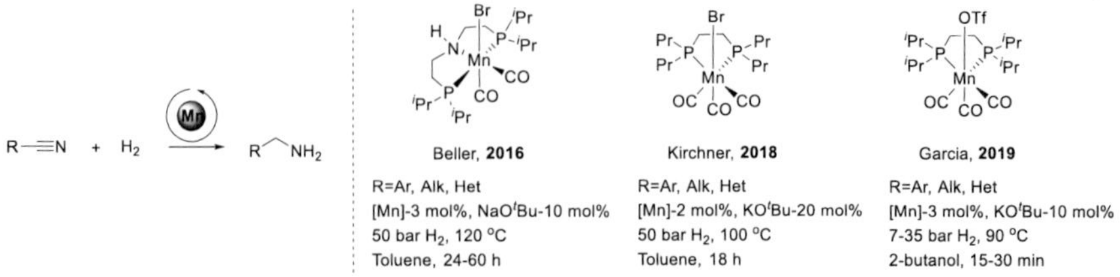

R≡N + H₂ —[Mn]→ R⁀NH₂

Beller, 2016
R=Ar, Alk, Het
[Mn]-3 mol%, NaOᵗBu-10 mol%
50 bar H₂, 120 °C
Toluene, 24-60 h

Kirchner, 2018
R=Ar, Alk, Het
[Mn]-2 mol%, KOᵗBu-20 mol%
50 bar H₂, 100 °C
Toluene, 18 h

Garcia, 2019
R=Ar, Alk, Het
[Mn]-3 mol%, KOᵗBu-10 mol%
7-35 bar H₂, 90 °C
2-butanol, 15-30 min

Scheme 7. Hydrogenation of nitriles.

Various substrates bearing electron-donating as well as electron-withdrawing functionalities were tolerated and provided primary amines in high yields. Rather harsh reaction conditions had to be applied compared to reduction of ketones and aldehydes.[38] The group of Kirchner reported biphosphine Mn(I) complex not only for the reduction of ketones and aldehydes but also for the reduction of nitriles. Thus, wide range of substituted nitriles was applied producing amines in high yields.[41] Later on, the group of Garcia reported another example of hydrogenation of nitriles using non-pincer manganese complex (Scheme 7).

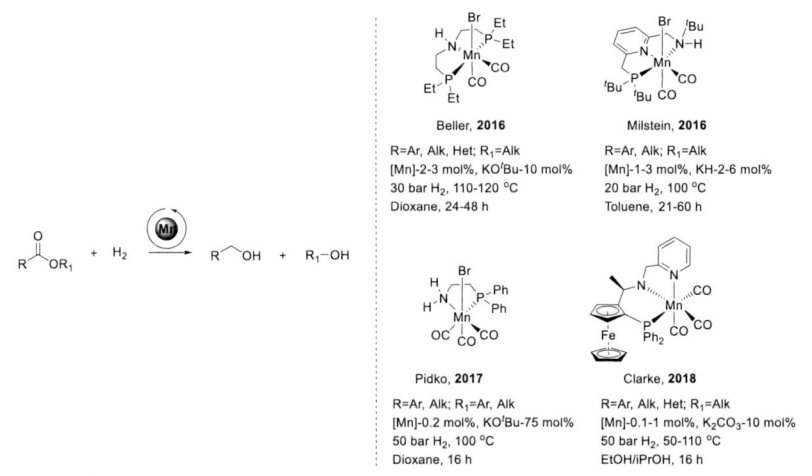

Scheme 8. Manganese-catalysed hydrogenation of esters.

The reaction proceeds very rapidly under mild reaction conditions providing desired amines in high yields.[44] A rapid development of manganese hydrogenation chemistry expanded also to hydrogenation of more challenging substrates, such as esters. The earliest example was reported by Beller and co-workers[45] in 2016. The application of a novel Mn-PNP complex resulted in high yields of the desired alcohols. Shortly after, Milstein and co-workers[46] presented Mn-PNN system which was successfully applied for the hydrogenation of a wide range of esters with high chemoselectivity towards ester group. Later on the group of Pidko[47] showed highly active Mn-PN complex for the hydrogenation of aromatic and aliphatic esters. In 2018, Clarke and co-workers[48] developed hydrogenation of enantiomerically pure esters without a significant loss of stereochemical integrity (Scheme 8). Hydrogenation of amides with an application of manganese based complexes was also explored, thus Beller and co-workers[49] developed Mn-PNN complex for a selective hydrogenation of amides to amines and alcohols whereas Milstein and co-workers[50] used lutidine based Mn-PNP complex with Lewis acid co-catalyst for deoxygenative hydrogenation of amides to amines (Scheme 9).

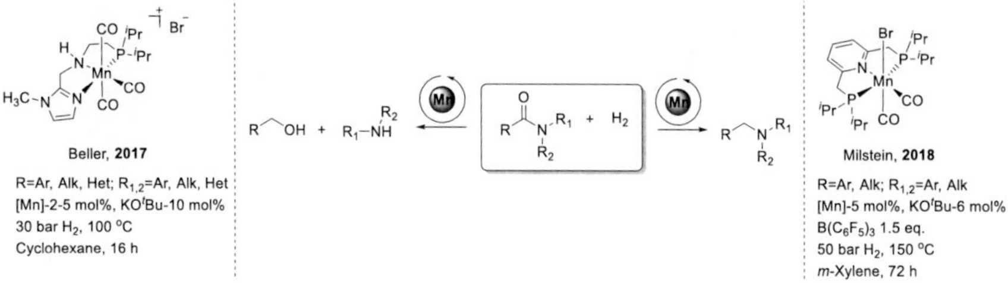

Scheme 9. Hydrogenation of amides catalysed by manganese complexes.

Dehydrogenation

Dehydrogenation reaction is the reverse of the hydrogenation and involves the removal of hydrogen from an organic molecule. This process is highly important because it can convert alcohols to ketones or aldehydes, saturated fats to unsaturated, amines to imines etc with a release of hydrogen which can be further used as a fuel. Additionally, green and sustainable procedures could be developed when intermediates of the initial dehydrogenation reaction take part in further reactions.[51] Ambitious goal is to transform alkanes, which are relatively inert and highly available, to olefins, which are reactive and more valuable. Dehydrogenation processes usually require elevated temperatures (above 500 °C), hence the development of catalytic systems, especially based on base metals, is highly required.

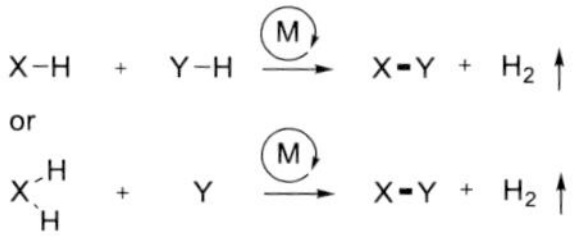

Figure 4. Metal catalysed dehydrogenation reactions.

In 2016, Beller and co-workers[52] presented the first manganese catalyst for the dehydrogenation of methanol/water mixtures (Scheme 10, left). The developed protocol showed great reactivity also for ethanol, paraformaldehyde and formic acid. Additionally, the catalyst demonstrates good long-term stability. At the same time, independently, the group of Boncella reported dehydrogenation and dehydration of formic acid using manganese pincer complex.[53] Later in 2017 Gauvin and co-workers[54] developed a catalytic system for acceptorless dehydrogenative oxidation of alcohols into their corresponding carboxylic acid salts (Scheme 10, right).

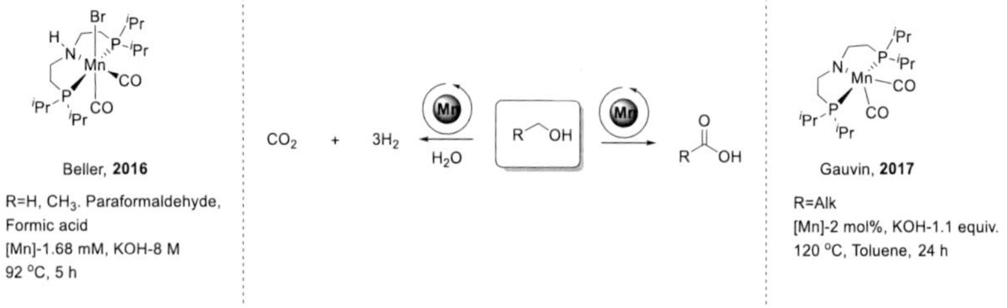

Beller, 2016
R=H, CH₃. Paraformaldehyde, Formic acid
[Mn]-1.68 mM, KOH-8 M
92 °C, 5 h

Gauvin, 2017
R=Alk
[Mn]-2 mol%, KOH-1.1 equiv.
120 °C, Toluene, 24 h

Scheme 10. Dehydrogenation of alcohols.

Acceptorless dehydrogenation (AD) reactions have major advantages. One of them is avoidance of stoichiometric amount of oxidant. Additionally hydrogen, a high-energy, clean fuel is released, while a minimum of waste is generated. Among dehydrogenation reactions few strategies stand out e. g. dehydrogenative coupling and dehydrogenative cyclisation reactions as presented below in Scheme 11. On the left AD with release of hydrogen gas is presented. The catalyst liberates H_2 from both starting compound and intermediate, following this strategy amides and esters can be formed by dehydrogenative coupling of primary alcohols with amines or with another alcohol. On the right the AD with H_2 and water release is shown, which can be represented by dehydrogenative coupling of alcohols with amines to produce imines that can be isolated or take part in further reactions.

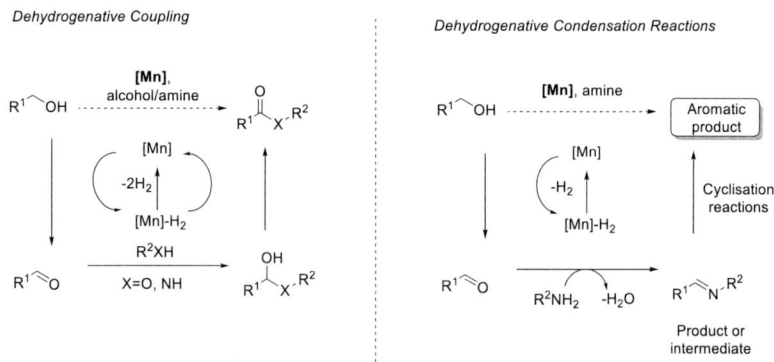

Scheme 11. Dehydrogenation reactions strategies.

Following the acceptorless dehydrogenative coupling (ADC) strategy, primary and secondary amines can be converted into the corresponding formamides as a result of methanol dehydrogenation. Thus, Milstein and co-workers reported N-formylation of primary and secondary amines using methanol[55] and later on expanded the procedure to other primary alcohols using newly developed Mn-PNN complex to achieve amides (Scheme 12).[56]

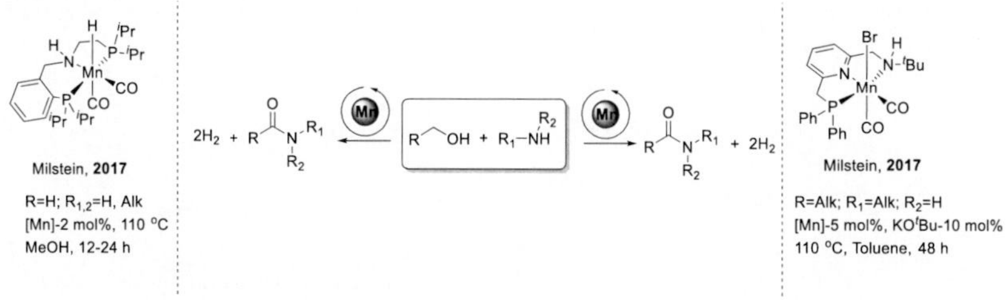

Scheme 12. Acceptorless dehydrogenative coupling of alcohols with amines.

Meanwhile Gauvin and co-workers presented the synthesis of esters starting from alcohols releasing two molecules of hydrogen.[57] Only 0.6 mol% of the manganese catalyst was sufficient for a successful reaction. Butane-1,4-diol was also applied under optimised reaction conditions leading to γ-butyrolactone as the only product. In the meantime, Milstein and co-workers reported a unique example for a synthesis of cyclic imides by dehydrogenative coupling of diols with amines.[58]

In contrast to dehydrogenative coupling reactions where the only by-product is hydrogen, in dehydrogenative condensation reactions, hydrogen and water are formed as by-products. In this regard, Milstein and co-workers presented manganese-catalysed coupling of alcohols and amines leading to imines.[59] A wide range of different alcohols and amines were tolerated giving desired imines in good to excellent yields. Later the Kirchner group independently reported another Mn-PNP complex to catalyse this reaction (Scheme 13).[60]

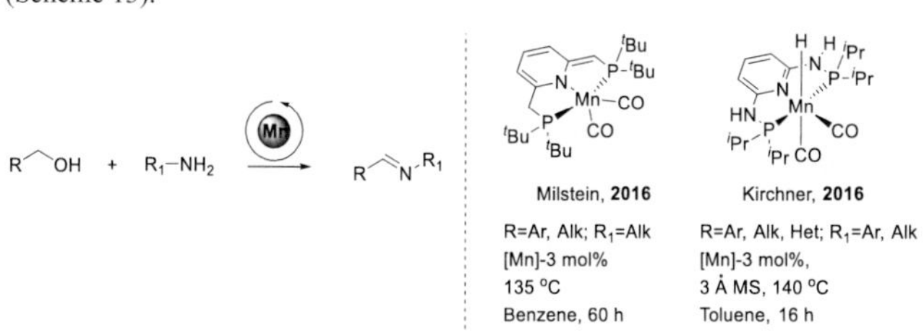

Scheme 13. Synthesis of imines by acceptorless dehydrogenation of alcohols and later condensation with amines.

Later on, Milstein and co-workers demonstrated, for the first time, direct deoxygenation of primary alcohols using base-metal catalyst.[61] A Mn-PNP complex catalysed reaction includes ADC of an alcohol with hydrazine giving hydrazone which undergoes Wolff-Kishner reduction liberating hydrogen. Primary benzylic and other aliphatic alcohols were applied providing desired products in moderate to excellent yields. Furthermore, Milstein and co-workers used the same manganese complex for olefination of nitriles by ADC of alcohols and nitriles providing a series of substituted acrylonitriles (Scheme 14).[62]

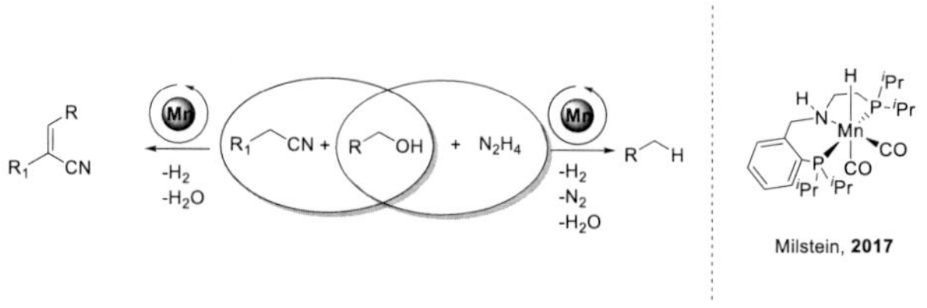

Scheme 14. Acceptorless dehydrogenative coupling of alcohols with nitriles (left) and with hydrazine (right).

Additionally, using acceptorless dehydrogenative coupling methodology a range of different heterocycles can be synthesised. Thus, Kirchner and co-workers applied Mn-PNP complex for a synthesis of quinolines starting from 2-aminobenzyl alcohols and secondary alcohols and synthesis of pyrimidines using benzamidine, primary and secondary alcohols.[63] Independently, the group of Kempe reported three and four component synthesis of pyrimidines from alcohols and amidines.[64] Moreover, Kempe an co-workers reported the Mn-catalysed synthesis of pyrroles from secondary alcohols and amino alcohols.[65] The reaction proceeds under mild reaction conditions, tolerates a wide range of functional groups and is easily scalable to more than 5 g of product.

Hydrogen autotransfer

A borrowing hydrogen or hydrogen autotransfer reaction is a valuable methodology and can be seen as an example of green chemistry. It occurs when, for example, an alcohol undergoes oxidative hydrogenation to become an aldehyde, which is further reacting with amine to give an imine which is consequently getting reduced to the amine liberating hydrogen which was gained through the dehydrogenation of the alcohol.[66–68] The method is highly atom economic as no additional activation for the alcohol is needed. Transfer hydrogenation (TH) reaction can be also discussed in the topic of the hydrogen autotransfer. TH is the addition of hydrogen to a molecule from any other source except of gaseous H_2. One of the most common hydrogen donors are alcohols (isopropanol, methanol), formic acid and Hantzsch ester. Transfer hydrogenation is a good alternative to a classic reduction with gaseous H_2 as it can be performed without special equipment, such as autoclaves. Manganese complexes were also studied in TH reactions. The first manganese complex which was found to catalyse TH of ketones was presented by Beller and co-workers.[69] Broad scope of substrates and high yields of the products were achieved. Independently, the Sortais group reported a highly efficient bidentate manganese complex for TH of ketones.[70] Meanwhile, Kirchner and co-workers expanded the procedure to asymmetric catalysis.[71] The reaction proceeds under room temperature giving high yields of the desired alcohols with *ee* from 20% to 86% (Scheme 15).

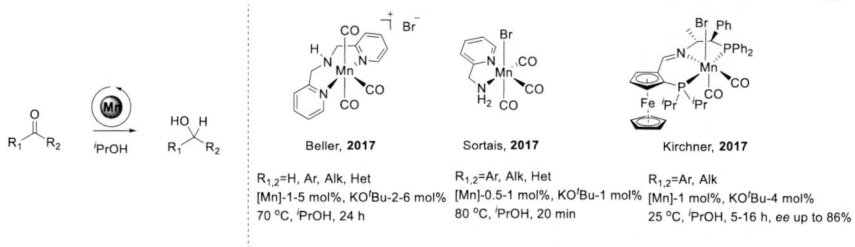

Scheme 15. Manganese-catalysed transfer hydrogenation of ketones.

In 2016, Beller and co-workers presented an elegant procedure for an alkylation of aromatic and heteroaromatic amines with alcohols using Mn-PNP complex for the first time.[72] The reaction proceeds under mild reaction conditions, various benzylic and aliphatic alcohols, including methanol, were applied and led to high yields of the desired alkylated amines with excellent chemoselectivity. Later on, the group of Beller presented a follow up work where they introduced lutidine based Mn-PNP complex particularly for methylation of amines with methanol.[73] Milder reaction conditions could be achieved giving methylated amines in high yields. Independently, Sortais and co-workers presented similar manganese complex which was able to catalyse methylation of amines.[74] Additionally, sulfonamides were successfully applied under more harsh conditions (Scheme 16).

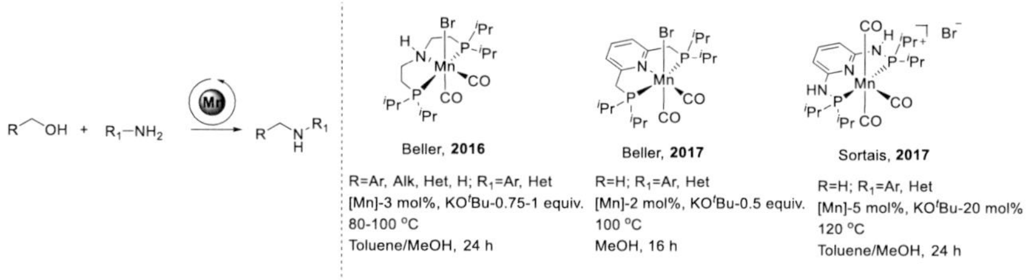

Scheme 16. Alkylation of amines with alcohols.

Manganese-catalysed hydrogen autotransfer reaction could also be applied for building C-C bonds. Thus, the Beller group reported alkylation of ketones using primary alcohols.[75] The reaction proceeds using low catalyst and base loading, but high temperature was required. Later, Liu and co-workers used the same manganese complex for upgrading ethanol to butanol.[76] The reaction, is known as Guerbet reaction and was performed for the first time using base-metal catalyst with selectivity and reactivity comparable to noble metals (Scheme 17).

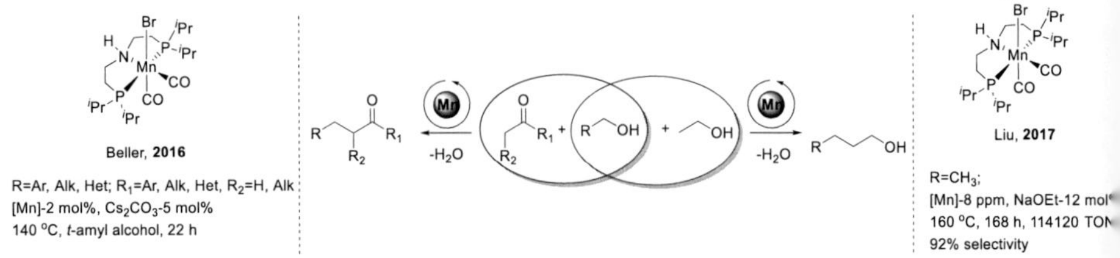

Beller, 2016
R=Ar, Alk, Het; R$_1$=Ar, Alk, Het, R$_2$=H, Alk
[Mn]-2 mol%, Cs$_2$CO$_3$-5 mol%
140 °C, *t*-amyl alcohol, 22 h

14
Liu, 2017
R=CH$_3$;
[Mn]-8 ppm, NaOEt-12 mol%
160 °C, 168 h, 114120 TON
92% selectivity

Scheme 17. C-C bond formation catalysed by Mn-PNP complex.

Chapter 2: Results and discussion

2.1 Manganese-catalysed hydrogenation of cyclic organic carbonates

2.1.1 Introduction

The catalytic hydrogenation of an abundant, safe and renewable CO_2 to methanol is very attractive and required due to "methanol economy". Among CO_2 hydrogenation products, methanol is most attractive because it is already one of the most important building blocks used today on a large scale to produce a variety of chemicals and products. It can be also directly used as alternative fuel for internal combustion and other engines, either in combination with gasoline or neat and in direct methanol fuel cells.[77] Direct CO_2 hydrogenation to methanol by homogeneous catalysis still remains a challenge, as a high activation energy barrier has to be overcome for the cleavage of the C-O bond.

It would be possible to avoid harsh conditions, while higher efficiency could be achieved by using the indirect pathway of CO_2 hydrogenation due better reactivity of CO_2-derived compounds than CO_2 itself. Outstanding results were achieved by the group of Milstein in the hydrogenation of CO_2-derived compounds on the example of methyl formate, dimethyl carbonate, ureas and a range of carbamates by dearomatized PNN-Ru(II) pincer complexes.[78] The hydrogenation of the cyclic carbonates to glycols and methanol using Ru(II) complex with general formula [(PNP)Ru-(CO)(Cl)] was reported by the group of K. Ding.[19]

The key for the sustainable future of industrial chemistry is the replacement of precious metal catalysts by the first-row, earth-abundant catalysts.[37] Surprisingly, only few works were devoted to the hydrogenation of CO_2/CO_2-derived compounds to methanol. In 2017, the group of Beller reported the hydrogenation of CO_2 by in situ generated Co-based catalyst.[79] Later, the group of G. K. Surya Prakash reported a manganese-catalysed sequential hydrogenation of CO_2 to methanol *via* formamides.[80] Thus, other catalytic systems are still desired for the mentioned transformation.

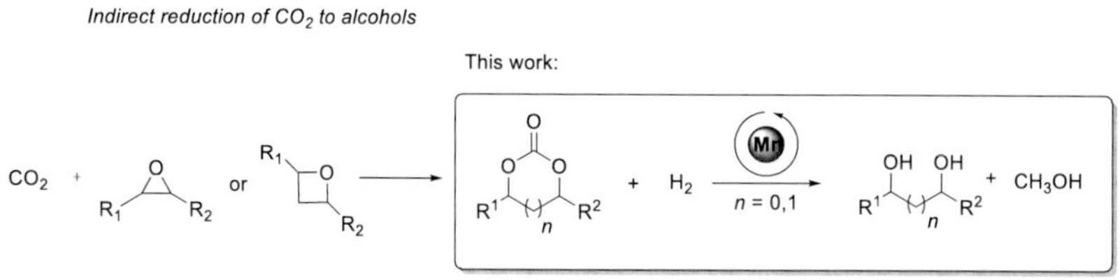

Scheme 18. Manganese-catalysed hydrogenation of cyclic organic carbonates to diols and methanol.

For our studies we have chosen the hydrogenation of cyclic organic carbonates (COCs) as this process is a 100% atom economical and substrates can be easily obtained from CO_2 and epoxides (Scheme 18).[81] Nowadays, 40 thousand tons of CO_2 are utilized annually to produce cyclic carbonates.[82] The hydrogenation of ethylene carbonate also produces ethylene glycol (EG) which is widely used in the chemical industry. On the other hand, Mn(I)-complexes showed to be active for hydrogenation of wide range of organic molecules.[83–88] Due to the high reactivity of Mn(I)-complexes for hydrogenation of esters, I wondered if they could be applied for the hydrogenation of cyclic carbonates with the same efficiency.

2.1.2 Optimization of reaction conditions

I started the investigations with the synthesis of a new air and moisture stable ^{Ph}PNN-Mn complex **Mn-1**, which is supported by a bench stable ^{Ph}PNN ligand **L1**. **Mn-1** was readily synthesised by treatment of **L1** with 1 equiv. of Mn(CO)$_5$Br as metal precursor in THF at 80 °C for 16 h. 84% of the pale-yellow complex was isolated and fully characterised by multinuclear NMR, IR, and mass spectrometry as well as single crystal X-ray diffraction study (Scheme 19).

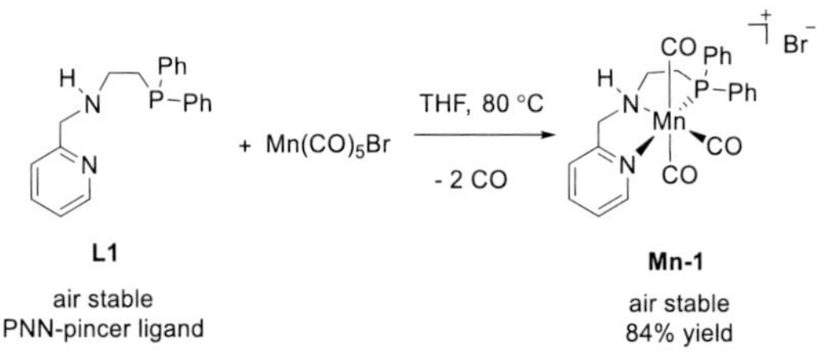

Scheme 19. Synthesis of complex **Mn-1**.

With the aim to find the right reaction conditions, ethylene carbonate was selected as a benchmark substrate for catalytic hydrogenation of COCs. To my delight, the new complex **Mn-1** showed excellent reactivity and provided EG in >99% yield and methanol in 92% yield when the reaction was performed at 50 bar hydrogen pressure in dioxane (Table 1, entry 1). Manganese complexes **Mn-2**[46] and **Mn-3**[47] previously studied in the hydrogenation of esters provided very poor reactivity towards the reduction of ethylene carbonate (Table 1, entries 2-3) highlighting the difficulty associated with carbonate reductions. Hence, I wondered whether a **Mn-4**[76] complex with more electron-donating aliphatic phosphines would result in better efficiency. Unfortunately, **Mn-4** led to lower catalytic activity (Table 1, entry 4). A control experiment proved that no reaction occurred in the absence of the metal catalyst (Table 1, entry 5) highlighting the possibility of developing low cost catalytic systems based on earth-abundant manganese catalysts. Subsequently, the influence of different solvents and bases was investigated. Employing K$_2$CO$_3$ as a base did not provide improve the results while Cs$_2$CO$_3$ was similar effective as KO^{t}Bu (Table 1, entries 6 and 7). Furthermore, testing various solvents such as THF, 2-methyl-THF and toluene did not lead to better results (Table 1, entries 8-10). Hence, dioxane was chosen as appropriate solvent. Importantly, further reduction of the catalyst loading to 0.5 mol% resulted in >99% of EG and 89% of methanol.

Table 1. Optimisation of the reaction conditions of manganese-catalysed hydrogenation of carbonates.[a]

entry	[Mn] (mol%)	base (mol%)	yield of EG (%)[b]	yield of MeOH (%)[b]
1	**Mn-1** (1)	KO^{t}Bu (2.5)	>99	92
2	**Mn-2** (1)	KO^{t}Bu (2.5)	20	17
3	**Mn-3** (1)	KO^{t}Bu (2.5)	08	02
4	**Mn-4** (1)	KO^{t}Bu (2.5)	66	56
5	-	KO^{t}Bu (2.5)	-	-
6	**Mn-1** (1)	K$_2$CO$_3$ (2.5)	>99	86
7	**Mn-1** (1)	Cs$_2$CO$_3$ (2.5)	>99	92
8[c]	**Mn-1** (1)	KO^{t}Bu (2.5)	>99	77
9[d]	**Mn-1** (1)	KO^{t}Bu (2.5)	90	80
10[e]	**Mn-1** (1)	KO^{t}Bu (2.5)	61	52
11	**Mn-1** (0.5)	KO^{t}Bu (1.25)	>99	89

[a] Reaction conditions: **1a** (1 mmol), [**Mn**], base in dioxane (0.25 M) at 140 °C under 50 bar of hydrogen for 16 h. [b] Determined by the GC analysis using m-xylene as internal standard. [c] Reaction in THF. [d] Reaction in 2-methyl-THF. [e] Reaction in toluene

2.1.3 Scope of substrates

In order to demonstrate the potential and applicability of the newly developed catalytic system, a range of COCs **1a-1m** were tested under the optimised reaction conditions (Table 2). An array of mono-substituted 5-membered 1,3-dioxolan-2-ones bearing different alkyl and aryl substituents such as Me, Et, *n*-Bu, Hex, *t*-Bu and Ph could be efficiently and selectively hydrogenated to the corresponding vicinal 1,2-diol and methanol in very good yields (Table 2, **2b-g**). The reaction tolerates the benzyloxymethyl and methoxymethyl derivatives **1h** and **1i** and the desired alcohols were produced in excellent yields (Table 2, **2h** and **2i**). Noteworthy, the disubstituted cyclic carbonate **1j** was successfully converted to methanol and 2,3-butylene glycol (**2j**) in very good yields. Under the same reaction conditions, different unsubstituted and substituted 6-membered COCs **1k-1m** were reacted in excellent yields to give methanol and the corresponding vicinal 1,3-diols.

Table 2. Manganese-catalysed hydrogenation of COCs.[a]

Substrate	Diol yield	Methanol yield
1b (R = H$_3$C)	99%	85%
1c (R = H$_3$C-CH$_2$)	94%	92%
1d (R = ^{n}Bu)	96%	88%
1e (R = H$_3$C-(CH$_2$)$_5$)	99%	95%
1f (R = ^{t}Bu)	99%	94%
1g (R = Ph)	99%	95%
1h (R = BnO-CH$_2$)	99%	95%
1i (R = H$_3$CO-CH$_2$)	99%	94%
1j (R = R' = CH$_3$)	99%	85%
1k (6-membered, unsubstituted)	99%	95%
1l (6-membered, H$_3$C)	99%	80%
1m[b] (6-membered, gem-diMe)	99%	99%

[a] Reaction conditions: **1** (1 mmol), **Mn-1** (1 mol%) and KO^{t}Bu (2.5 mol%) in dioxane (0.25 M) at 140 °C under 50 bar of hydrogen for 16 h. Isolated yields of diol products except for substrates **1b**, **1c** and **1k** (GC yield using m-xylene as internal standard). Yield of methanol is determined by the GC analysis using m-xylene as internal standard. [b] **Mn-1** (1.2 mol%) and KO^{t}Bu (3 mol%).

Reaction scheme: COC **1a-m** + H$_2$ (50 bar), $n = 0, 1$ → with **Mn-1** 1 mol%, KO^{t}Bu 2.5 mol%, 140 °C, 16 h, dioxane 0.25 M → diol **2a-m** + CH$_3$OH.

Additionally, polycarbonates, which known to be produced from CO_2,[89] were found to be active in the hydrogenative depolymerization process by using 2 mol% of the catalyst **Mn-1**. Hence, poly(propylene carbonate) (PPC) was efficiently hydrogenated to the corresponding propane-1,2-diol and methanol providing 99% and 87% yields respectively (Scheme 20). Thus, the above-mentioned catalytic system provides the possibility to recycle polycarbonate waste. Furthermore, formamides could be selectively reduced to amines and methanol using only 0.5 mol% of the catalyst highlighting high versatility of the developed system (Scheme 21).

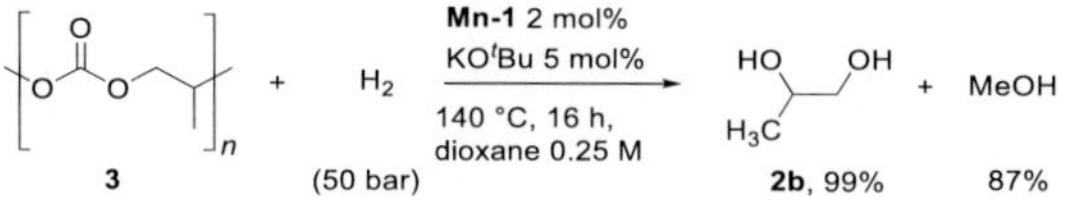

Scheme 20. Mn-1 catalysed hydrogenation of polycarbonates.

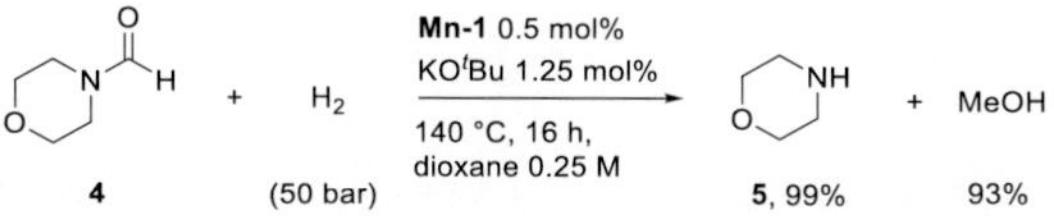

Scheme 21. Application of **Mn-1** for the hydrogenation of formamides.

2.1.4 Mechanistic investigations

Deuterium labelling studies were conducted to get more inside in the reaction mechanism. Due to lower reactivity of D_2 compared to H_2 higher catalyst loading was used to perform the reaction. Application of ethylene carbonate in the deuteration reaction resulted in the formation of methanol with >95% deuterium incorporation in the methyl group, whereas ethylene glycol was formed with no deuterium incorporation (Scheme 22a). This indicates a much faster deuteration of ethylene carbonate compared to dehydrogenative deuteration of ethylene glycol. Achieved results show great applicability of the developed system in synthesis of widely used deuterated methanol. Additionally, the treatment of EG with deuterium was performed resulting in significant deuterium substitution in the carbon atoms which points out the potential of the developed catalyst in the dehydrogenative deuteration process (Scheme 22b).

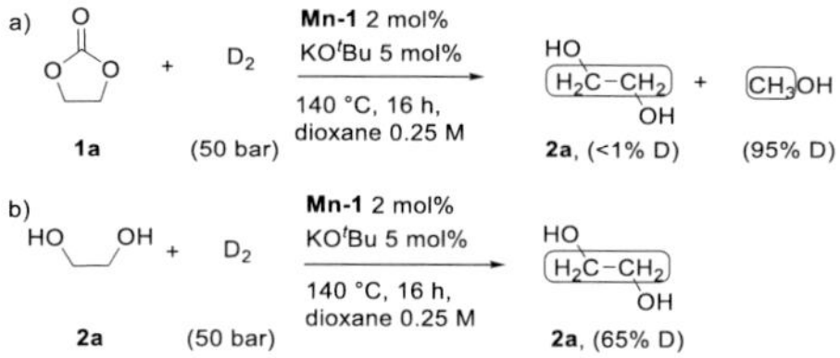

Scheme 22. Deuterium-labelling experiments.[1]

Additionally, the hydrogenation of the intermediates in the studied reaction was carried out. Treating paraformaldehyde with 2 mol% of the catalyst lead to the formation of 83% of methanol which proves that formaldehyde is involved in the catalytic cycle (Scheme 23).

---

[1] The experiments were performed by Dr. Y. Lebedev

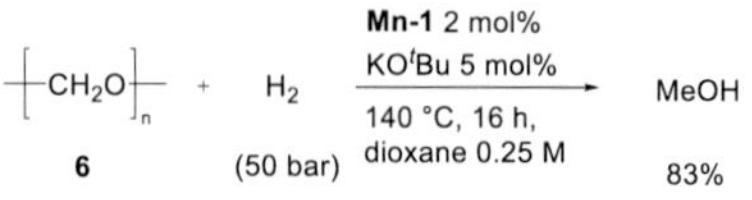

Scheme 23. Manganese-catalysed hydrogenation of paraformaldehyde.

Hydrogenation of hydroxyformates **1f'** and **1f''** which are possible intermediates in this reaction was carried out in one pot with starting cyclic carbonate **1f** in order to identify rate determining step in this competition experiment. The mixture was applied to optimal reaction conditions with shortened time of 8 h leading to a full conversion of intermediates **1f'** and **1f''** along with a recovery of 64% of the cyclic organic carbonate **1f**. The result shows that the hydrogenation of starting carbonate is more energetically uphill compared to their corresponding formats (Scheme 24).

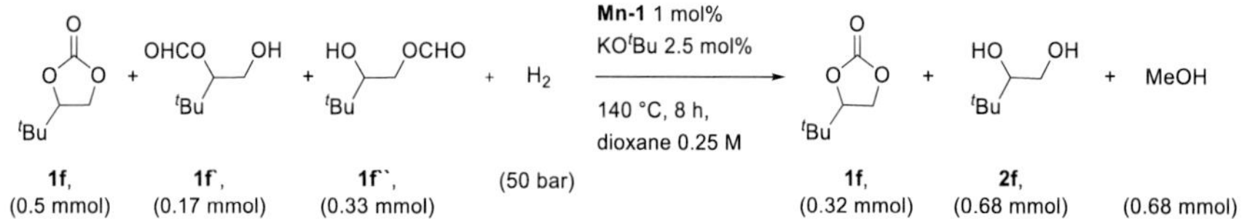

Scheme 24. Competitive hydrogenation of organic carbonates with formats.

2.1.6 Proposed reaction mechanism

The mechanism of the Mn(I)-PNN-catalysed hydrogenation of cyclic organic carbonates was also investigated by DFT calculations.[2][90] The overall hydrogenation of ethylene carbonate to ethylene glycol and methanol can be separated into three independent C=O hydrogenation events, each with a corresponding catalytic cycle. The first one is the hydrogenation of ethylene carbonate (**1a**) into 2-hydroxyethyl formate, the second one is the hydrogenation of 2-hydroxyethyl formate into ethylene glycol (**2a**) and formaldehyde, and the third one is the hydrogenation of formaldehyde into CH$_3$OH. In the following, the first hydrogenation cycle only will be discussed (Figure 5). The other two catalytic cycles are composed of similar steps. Calculations were performed using the most active **Mn-1** catalyst. During the initiation process concerning the H$_2$ addition to the Mn active site (steps **A** to **C**, where **A** is a 16-electron species), H$_2$ coordination to Mn is a non-spontaneous process demanding 14.6 kcal mol^{-1} at the M06/TZVP//wB97XD/SVP(H,C,N,O,P)-TZVP(Mn) level of theory in 1,4-dioxane as the solvent and relative to **A**. The heterolytic cleavage of H$_2$, *via* transition state **B-C** (at 21.3 kcal mol^{-1} relative to **A**), leads to the hydrogenation of the catalytic species by hydride addition to the Mn centre and protonation of the nonpyridinic N atom. The resulting [Mn]-H$_2$ species **C** (zero energy reference) promotes the reduction of the C=O bond of all substrates in the three catalytic cycles. The C=O hydrogenation step is characterised

---

[2] DFT calculations were performed by Dr. L. M. Azofra

by two main mechanistic events. On the one hand, the nucleophilic character of the hydride on Mn promotes hydride transfer from Mn to the C(sp^2) atom of the substrate (steps **D** to **E**). On the other hand, in agreement with the literature, proton transfer is achieved through cleavage of the H-H bond of a H$_2$ molecule η^2-coordinated to the metal (steps **F** to **H**)[42,91]. This prevents formation of the 16-electron species **A** by deprotonation of the N-H functionality. In more detail, the first step is the hydride transfer from the catalyst to the substrate *via* transition state **D-E** at 16.4 kcal mol^{-1}, leading to the Mn alkoxide intermediate **E**, at -3.9 kcal mol^{-1}. It can be seen that as a consequence of this hydride transfer, 1,3-dioxolan-2-olate is transformed into 2-(formyloxy)ethan-1-olate by charge reorganisation in the substrate. The second step is proton transfer from a coordinated H$_2$ molecule, releasing the product. In the presence of an excess of methanol, which is expected as the reaction evolves, metathesis of **E** with CH$_3$OH can lead to the methoxide species **F**. Coordination of H$_2$ leads to **G** and triggers rehydration of the Mn active site and proton transfer to methanolate *via* transition state **G-H** (at 22.1 kcal mol^{-1}). This entails the regeneration of CH$_3$OH as well as the catalytic species. Based on calculations, the rate-determining transition state along the cycle is this proton transfer, with an overall activation barrier of 25.0 kcal mol^{-1} from **E** to **G-H**. Of course, this proton transfer can also occur without involving initial metathesis of CH$_3$OH with **E**, with very similar reaction steps involving the Mn alkoxide bond of **E**. The impact of the phosphine substituents was analysed by comparing **Mn-1** with catalysts presenting aliphatic and bulky ^{t}Bu substituents on the P atom and the less sterically impeded ^{i}Pr. Hydrogenation during the initiation process of both alkylsubstituted catalysts, *via* transition state **B-C**, exhibits larger barriers of 25.3 and 23.7 kcal mol^{-1} compared to **Mn-1** (21.3 kcal mol^{-1}) and relative to **A**. In this sense, the DFT predictions are in agreement with experiments suggesting that the ^{Ph}PNN ligand confers to the manganese catalyst superior behaviour over those functionalised with aliphatic phosphines.

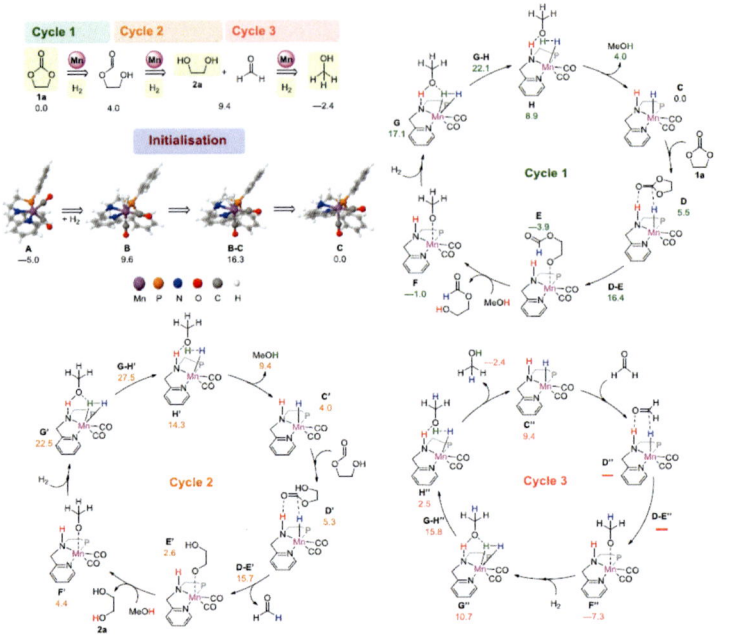

Figure 5. Proposed reaction mechanism for the three-cascade hydrogenation of ethylene carbonate into methanol plus EG catalysed by **Mn-1** (**P** = PPh$_2$). Steps **A** to **C** refer to the initialisation by catalyst hydrogenation, while steps **D** to **H** concern the substrate hydrogenation and catalyst regeneration by MeOH and η^2-H$_2$ assistance. Calculated Gibbs free reaction and activation energies, at 140 °C and 50 bar reaction conditions, are shown in kcal mol^{-1} at the M06/TZVP//ωB97XD/SVP(H,C,N,O,P)-TZVP(Mn) computational level in 1,4-dioxane as solvent. Note: H–N–Mn–H species (**C**, Cycle 1) is relative zero in energy. Cycle 2 and 3 represent the hydrogenation of the intermediates.

During the exploration of the potential energy surface (PES), the existence of a high-energy minimum was noticed, labelled as **E*** in Figure 6, as result of the hydride transfer (**D-E** TS). This minimum **E*** is, for the case of Cycle 1 (hydrogenation in ethylene carbonate catalysed by **Mn-1**), 2.3 kcal mol^{-1} more favoured in electronic energy than the immediate precedent **D-E** TS, however, after single-point refinement and including entropy and thermal corrections, this is computed as 1.4 kcal mol^{-1} less stable in Gibbs free energy at 140 °C and 50 bar reaction conditions. For this reason, minimum **E*** was omitted from the main analysis.

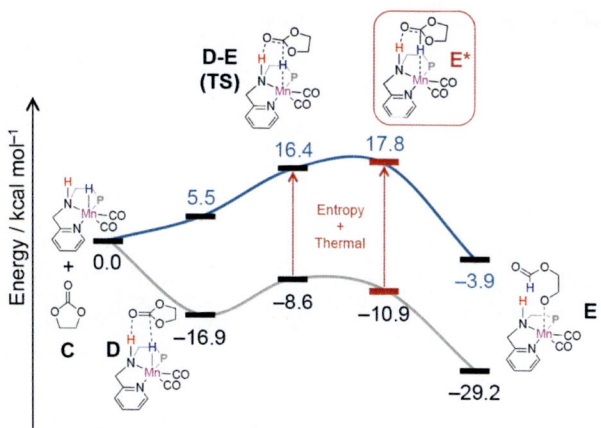

Figure 6. Energy diagram for **C** to **E** steps in Cycle 1 (hydrogenation in ethylene carbonate catalysed by **Mn-1**). Grey line refers to electronic energies at the ωB97XD/SVP(H,C,N,O,P)-TZVP(Mn) level. Blue line refers to Gibbs free energies after single-point refinement at M06/TZVP level. All energy values are shown in kcal mol^{-1} at 140 °C and 50 bar reaction conditions in 1,4-dioxane as solvent. Note: for the purposes of this graph, all energy values have been referred (0 kcal mol$_{-1}$) to isolated [Mn]-H$_2$ plus ethylene carbonate (**C**) substrates.

2.1.7 Summary

In the described project the hydrogenation of the CO$_2$-derived cyclic organic carbonates to alcohols using a homogenous base metal complex was studied. The reaction is catalysed by air and moisture stable Mn-PNN complex, proceeds with high efficiency and selectivity under mild conditions, without generation of any waste or side products. A range of 5-and 6-membered cyclic organic carbonates were applied to achieve methanol and vicinal diols in very good to quantitate yields, providing indirect route for the production of methanol from CO$_2$. Due to increased amount of CO$_2$ in the atmosphere the developed catalytic system could become a current alternative to the hydrolysis in omega process. Additionally, the method was successfully applied for the recycling of polycarbonates producing corresponding diols and methanol in high yields. The reaction proceeds *via* metal-ligand cooperative catalysis, which was characterized and proved by density functional theory and D-labelling experiment. From the computational studies I can conclude that the rate determining step of the whole catalytic process is the hydride transfer from manganese to sp^2 carbon of the cyclic organic carbonate.

Independently parallel to this studies Milstein and co-workers[92] and Leitner and co-workers[93] reported hydrogenation of organic carbonates to alcohols using manganese pincer complexes.

2.2 Diastereoselective amination of racemic alcohols

2.2.1 Introduction

The preparation of chiral amines has been a subject of great importance since the amine functionality is present in many pharmaceuticals, natural products and biologically active compounds.[94–96]

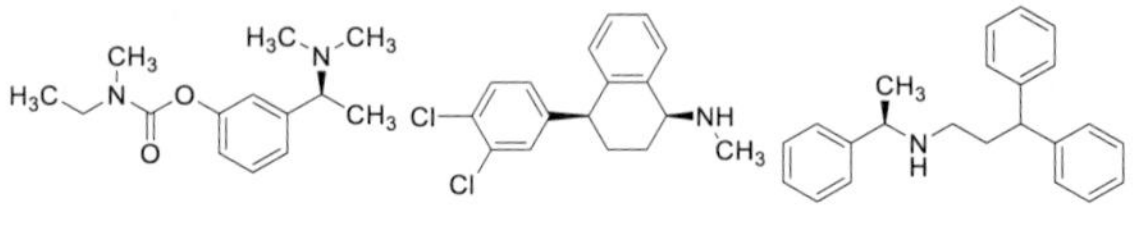

Figure 7. Examples of drugs with α-chiral amine moieties.

A straightforward approach to access chiral amines would be asymmetric hydrogenation of imines. [95,96,97] Unfortunately, this approach still has major limitations related to substrate scope and activity of the catalysts. Recently the hydrogen autotransfer (HA) strategy has gained increased attention mainly due to its synthetic importance as a powerful environmentally benign method for the construction of C-C and C-N bonds. In this regard, progress has been made in transition-metal catalysed *C*-alkylations and *N*-alkylations with non-activated alcohols. Mainly noble metal catalysts were used to produce achiral or racemic products.[51,66,98,99,100,101] The direct asymmetric amination of alcohols to produce chiral aniline,[102,103] amino alcohols,[104] oxazolidinones,[105] and hydrazones[106] has been demonstrated using ruthenium and iridium catalysis. In contrast, the direct synthesis of chiral primary amines from racemic alcohols is significantly more challenging. Until now only one protocol employing a ruthenium catalyst was reported.[107] However, this noble metal catalysed method is mainly limited to methyl substituted chiral amines. Therefore, the development of new catalytic systems, ideally based on a nonprecious metal catalyst, which can transform a wide range of racemic alcohols to chiral primary amines is an attractive goal. Besides, the replacement of the toxic noble metal catalysis by first row metals including Fe, Co or Mn has economic and ecological benefits.[37] Encouraged by a high reactivity and versatility of manganese pincer complexes[83–88] I was interested to employ manganese catalysts for the synthesis of chiral primary amines from highly accessible racemic alcohols. Since its discovery Ellman's sulfinamide[108] has become widely used for the synthesis of α-chiral primary amines, with many industrially relevant applications.[108,109] The idea of the project was to employ Ellman`s reagent for the reaction with racemic alcohols avoiding at the same time three steps synthesis, i.e, stoichiometric oxidation, condensation using a titanium reagent and stoichiometric reduction (Scheme 25).

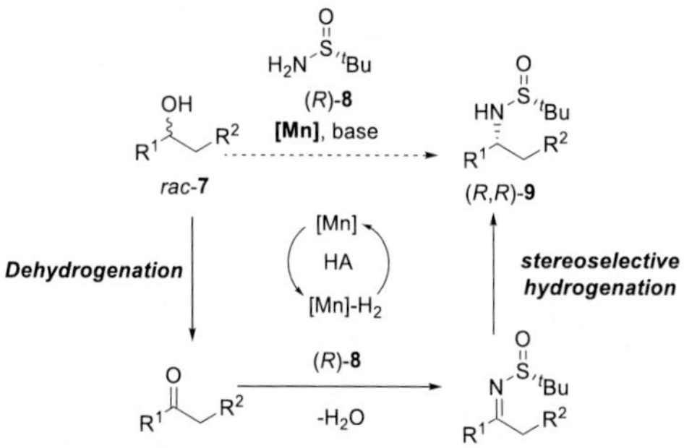

Scheme 25. Manganese-catalysed diastereoselective hydrogen autotransfer.

2.2.2 Optimization of reaction conditions

Preliminary studies involved the use of 1-phenylethanol (**7a**) and the sulfinamide (*R*)-**8** in the presence of **Mn-1** (2.5 mol%) and KO*t*Bu (10 mol%) at 140°C in toluene leading to the desired product in 33% NMR-yield with very good diastereomeric ratio 98:2 (Table 3, entry 1). When the complex **Mn-4** bearing aliphatic phosphine was tested, significant decrease in the reactivity was observed giving only 9% of the desired amine (Table 3, entry 2). Next, the reactivity of pincer manganese complexes bearing different types of PNP ligands were evaluated. In more details, the lutidine-based complex **Mn-5**,[110] aliphatic NH-bridged complex **Mn-6**[111] and triazine-based complex **Mn-7**[39] showed unsatisfactory results (Table 3, entries 3-5). Based on these results, the reaction conditions were further optimised with **Mn-1**. Next, different solvents were explored. The polar aprotic solvents such as dioxane and 2-methyl-THF showed lower performance compared to the toluene (Table 3, entries 6-7). A slightly better result (39% yield) was obtained when the reaction was performed in the polar protic *t*-amyl alcohol (Table 3, entry 8). Next, different bases were evaluated for this reaction. Potassium hydroxide has shown similar results as the KO*t*Bu while potassium carbonate was leading to significantly lower catalytic activity (Table 3, entries 9-10). Later the use of Cs$_2$CO$_3$ in combination with *t*-amyl alcohol was found to be the optimal combination for this reaction (Table 3, entries 11-12). Finally, increasing the catalyst loading to 5 mol% resulted in 85% isolated yield of **9a** with excellent diastereoselectivity (Table 3, entry 13).

Table 3. Optimization of the reaction conditions for the diastereoselective amination of **7a**.[3] [a]

Ph-CH(OH)-CH₃ + H₂N-S(O)₂-tBu → HN(S(O)₂tBu)-CH(Ph)-CH₃
7a   (R$_s$)-8   9a
[Mn], base, 140 °C, 16 h

Mn-1, Mn-4, Mn-5, Mn-6, Mn-7

entry	[Mn] (mol%)	base (mol%)	solvent	yield of 9a (%)[b]	dr
1	**Mn-1** (2.5)	KO^{t}Bu (10)	toluene	33	98:02
2	**Mn-4** (2.5)	KO^{t}Bu (10)	toluene	9	nd
3	**Mn-5** (2.5)	KO^{t}Bu (10)	toluene	<5	nd
4	**Mn-6** (2.5)	KO^{t}Bu (10)	toluene	10	nd
5	**Mn-7** (2.5)	KO^{t}Bu (10)	toluene	<5	nd
6	**Mn-1** (2.5)	KO^{t}Bu (10)	dioxane	27	98:02
7	**Mn-1** (2.5)	KO^{t}Bu (10)	2-methyl-THF	14	nd
8	**Mn-1** (2.5)	KO^{t}Bu (10)	TAA	39	99:01
9	**Mn-1** (2.5)	KOH (10)	TAA	41	99:01
10	**Mn-1** (2.5)	K$_2$CO$_3$ (10)	TAA	18	97:03
11	**Mn-1** (2.5)	Cs$_2$CO$_3$ (10)	TAA	45	99:01
12	**Mn-1** (2.5)	Cs$_2$CO$_3$ (5)	TAA	48	99:01
13	**Mn-1** (5)	Cs$_2$CO$_3$ (10)	TAA	**85**[c]	**99:01**

[a] Reaction conditions: **7a** (0.75 mmol), **8** (0.5 mmol), [Mn] and base in toluene (0.5 M) at 140 °C in a glass tube under an inert atmosphere for 16 h. [b] Yields were determined by the ¹H NMR analysis of the crude reaction mixture using mesitylene as an internal standard. [c] Isolated yield. TAA is *t*-amyl alcohol

2.2.3 Scope of substrates

Having optimised reaction conditions, the variability and the applicability of the asymmetric hydrogen autotransfer reaction were investigated (Table 4). Initially, different racemic benzylic alcohols bearing *β*-methyl group were explored. The alcohols **7a-7i** bearing different electron donating and electron withdrawing substituents were applied leading to desired products **9a-9i** without significant effect on the reactivity or the stereochemical outcome. Moreover, all products were isolated in very good yields and with

---

[3] Optimisation of reaction conditions was performed by M. A. Tran. The results were used to obtain Master of Science degree at RWTH Aachen University.

excellent optical purity. Similarly, the naphthyl substituted sulfinamide **9j** was afforded in very good yield and stereoselectivity. Importantly, some of these amines are key intermediates in the synthesis of pharmaceuticals and bioactive molecules. For example, Carpropamid,[112] an agriculture fungicide, is prepared from (R,R_s)-**9b**, whereas the Alzheimer's and Parkinson's drug Rivastigmine is produced using the sulfinylamine (S,S_s)-**9g**.[113] Notably, the traditional synthesis of α-chiral amines bearing a β-methyl group which involves the addition of MeLi to *N-tert*-butylsulfinylaldimines suffers from the low diastereocontrol even at low temperature.[108,109] The established catalytic system was found to not be limited to the methyl substituted alcohols. Thus, 1- tetralol (**7k**) was converted to the desired product in 84% yield and 94:06 *dr*. This product is highly relevant as it is used in the synthesis of diverse of bio-related compounds.[94,114,115] The ethyl substituted alcohol **7l** and the more challenging butyl substituted alcohol **7m** were well tolerated using the presented catalytic system. Next, more demanding non-benzylic alcohols **7n**-**7q** were investigated. The chiral amines bearing cyclohexyl substituent **9n** and cyclopropyl substituent **9p** were produced in very good diastereomeric ratio. Interestingly, the Amphetamine, which is used in the treatment of attention deficit hyperactivity disorder can be obtained from the *rac*-**7q** in 78% yield and good selectivity. Heterocycles-containing alcohols were tolerated and afforded the chiral amines **9r**-**9u** in very good yields with high diastereoselectivity. Despite the importance of the optically pure **9r** and **9s** in the synthesis of HIV protease inhibitors, the asymmetric synthesis of the corresponding primary amines is not yet reported.[116]

Although the scope of substrates is quite broad, many other substrates were inactive or decomposed under applied reaction conditions. Thus, *N*-substituted heterocycles, except of compound **7u**, were unsuitable for described transformation. Furane and benzofurane containing substrates were also inactive. *Ortho*-substituted compounds were not reactive most probably due to a steric hindrance. A full overview of the reaction limitations is presented in the Scheme 26.

Table 4. Manganese-catalysed asymmetric amination of *sec*-alcohols.[4][a]

Entry	Yield, dr
(R,R$_s$)-**9b**	80%, 98:02 dr
(R,R$_s$)-**9c**	65%, 98:02 dr
(R,R$_s$)-**9d**[b]	74%, 97:03 dr
(R,R$_s$)-**9e**	87%, 97:03 dr
(R,R$_s$)-**9f**	85%, 99:01 dr
(R,R$_s$)-**9g**	83%, 98:02 dr
(R,R$_s$)-**9h**	95%, 98:02 dr
(R,R$_s$)-**9i**[c]	70%, 96:04 dr
(R,R$_s$)-**9j**	76%, 97:03 dr
(R,R$_s$)-**9k**[c]	84%, 94:06 dr
(R,R$_s$)-**9l**[b]	80%, 94:06 dr
(R,R$_s$)-**9m**[b,d]	86%, 97:03 dr
(R,R$_s$)-**9n**	55%, 93:07 dr
(R,R$_s$)-**9o**[b,d]	74%, 82:18 dr
(R,R$_s$)-**3p**[c]	89%, 93:07 dr
(R,R$_s$)-**9q**	78%, 83:17 dr
(R,R$_s$)-**9r**[b,d]	86%, 96:04 dr
(R,R$_s$)-**9s**	90%, 92:08 dr
(R,R$_s$)-**9t**	83%, 80:20 dr
(R,R$_s$)-**9u**[b,d]	65%, 95:05 dr

[a] Reaction conditions: **7** (0.75 mmol), (R$_s$)-**8a** (0.5 mmol), **Mn-1** (0.025 mmol) and Cs$_2$CO$_3$ (0.05 mmol) in *t*-amyl alcohol (0.5 M) were stirred at 140 °C (aluminum block), for 16 h in a glass tube under argon. Yields after column chromatography are given. [b] **Mn-1** (0.05 mmol), Cs$_2$CO$_3$ (0.1 mmol). [c] 48 h. [d] **7** (1 mmol).

---

[4] Scope of substrates was performed in cooperation with M. A. Tran. The results were used to obtain Master of Science degree at RWTH Aachen University.

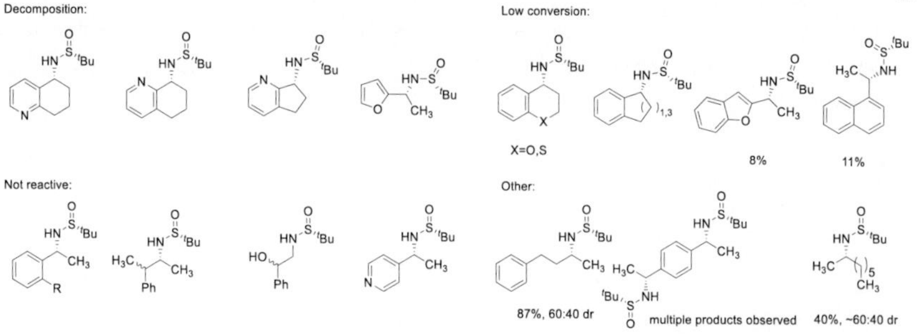

Scheme 26. Scope limitations.

2.2.4 Mechanistic studies

In order to gain more insight into the reaction mechanism, deuterium-labeling experiments were carried out (Scheme 27). To avoid the D/H exchange between the solvent and the substrates, the reactions were carried out in toluene instead of the *t*-amyl alcohol. An experiment with [D$_1$]-1-phenylethanol provided the desired product **9a'** in 27% yield with 73% deuteration at the α-position (Scheme 27a). As complementary experiment, when 1-phenylethanol-OD was subjected to the coupling reaction, high deuterium incorporation only at the methyl group was observed (Scheme 27b). The observed deuterium unscrambling between the carbon and the heteroatoms support monohydride mechanism and highlights the involvement of both the metal and non-innocent ligand.[117–119,120] The deuterium incorporation on the methyl group and not the NH-moiety can be explained by the prior equilibration of the imine and enamine tautomers of the sulfinamide as well as the keto-enol form of the acetophenone. Furthermore, in the transfer hydrogenation of the imine **4e** using [D$_1$]-1-phenylethanol as a hydrogen donor, the desired product **9e'** was produced in 66% D at α-position along with the formation of acetophenone, which supports the hydrogen transfer pathway (Scheme 27c). Finally, when a competitive experiment between **7a** and [D$_1$]**7a** was performed the desired product was obtained in 78% yield with 21% deuterium incorporation which eliminates the possibility of the nucleophilic substitution pathway (Scheme 27d).[121]

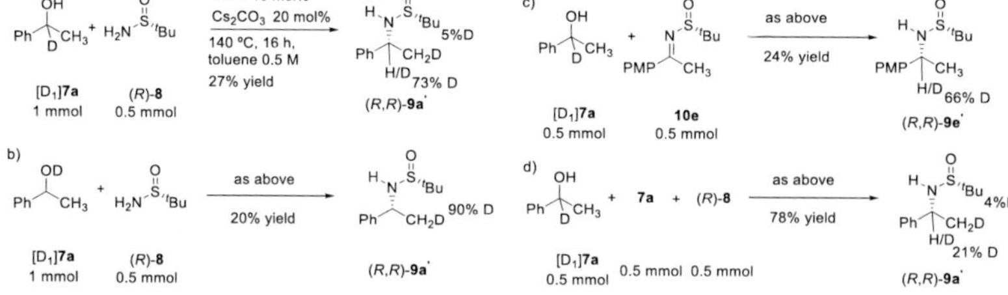

Scheme 27. Deuterium-labelling experiments.

Additionally, kinetic studies were performed using optimised reaction conditions. After a selected time the reaction mixtures were analysed by NMR and the progress of the reaction is shown in Figure 8. Already after 4 hours a significant amount (70%) of the product is formed. In the next 12 hours the yield of the desired product increases to 93%. During the reaction I observed the formation of less than 10% of acetophenone, only to the end of the reaction the amount of the acetophenone increases to 14% (Figure 8).

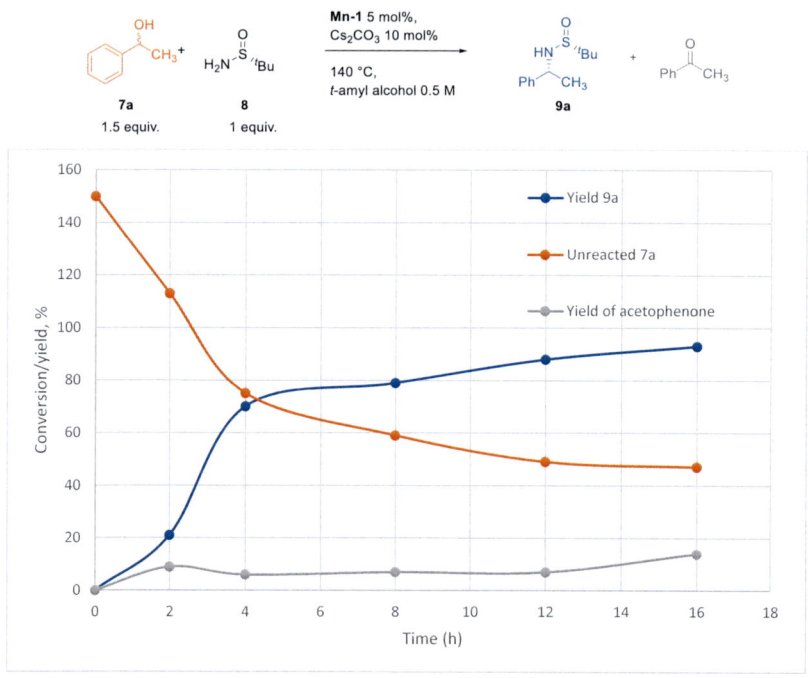

Figure 8. Kinetic profile for the diastereoselective amination of 1-phenylethan-1-ol **7a**.

2.2.5 Proposed reaction mechanism

The proposed reaction mechanism is shown in Scheme 28. Initially, proton transfer from the alcohol to the activated manganese species **A** occurs which leads to intermediate **B**. Following hydride transfer takes place with a release of the ketone and production of the hydrogenated catalyst **C**. Next, the condensation reaction between the *in situ* generated ketone and (*R*)-**8** takes place leading to the formation of an imine which might be in equilibrium with the corresponding enamine tautomer. The taken hydrogen stored in catalyst **C** is then transferred to C=N bond in two steps, where the hydride transfer and the protonation of the substrate occurs. Finally, the active manganese catalyst **A** is regenerated when the desired product is released. The proton transfer may also occur from the alcohol to give the desired product and the intermediate **B** without the re-generation of the 16e species **A**.[122]

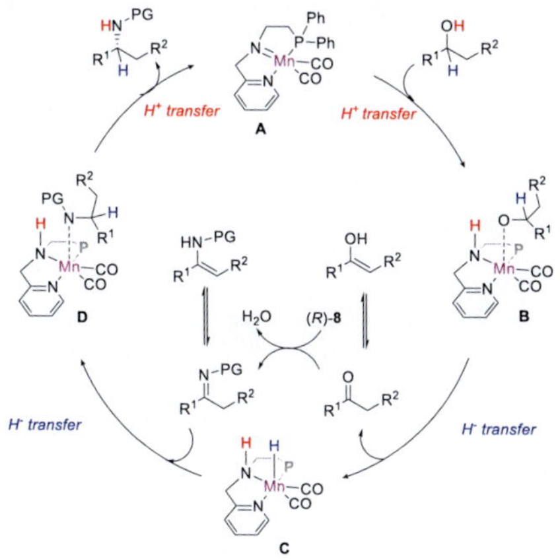

Scheme 28. Proposed reaction mechanism.

2.2.6 Summary

In conclusion, the first base metal catalysed asymmetric amination of racemic alcohols using hydrogen autotransfer strategy was developed. The reaction is catalysed by a bench-stable homogenous manganese complex which can be scaled-up. Notably, this environmentally benign, atom economic protocol uses inexpensive earth abundant metal and readily available substrates producing water as sole by-product. Thus, the presented catalytic system can serve as a basis for further application in the synthesis of relevant optically pure α-chiral amines and heterocycles.

2.3 Hydrogenation of alkynes and alkenes catalysed by manganese pincer complexes

2.3.1 Introduction

The selective semihydrogenation of alkynes to alkenes is a valuable catalytic process.[123,124] It is particularly important as it leads to building blocks relevant for the synthesis of pharmaceuticals, agrochemicals and natural products, which include double bond in defined *(E)*- or *(Z)*- configuration.[125] Different approaches were developed to prepare *(Z)*-alkenes.[126] Among them the application of Lindlar`s catalyst is the most popular and widely used method to form *(Z)*-alkenes from alkynes.[127] However, the reaction has disadvantages such as the toxicity of lead additives, often observed isomerization of the achieved *(Z)*- to *(E)*-isomer as well as a shift of the double bond. Thus, the development of cost-effective, well-defined, efficient and environmentally friendly catalytic systems for the selective conversion of internal alkynes to *(Z)*-alkenes is required (Scheme 29).

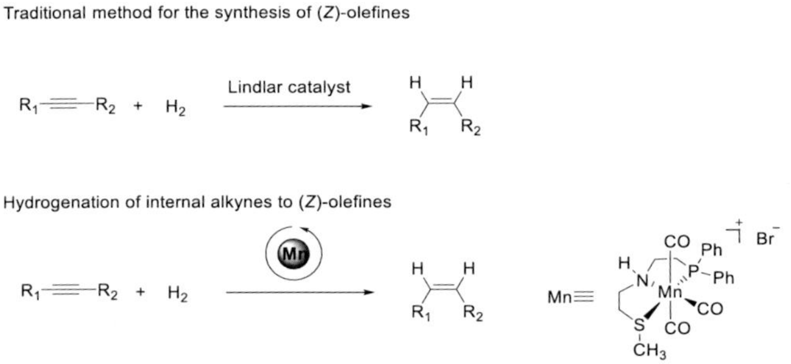

Scheme 29. Manganese-catalysed hydrogenation of alkynes to *(Z)*-alkenes.

The semireduction of alkynes to alkenes using hydrogen as a reducing agent is the most efficient and atom economic approach. Concerning homogeneous catalysis most of the known procedures rely on using alternative hydrogen donors, such as formic acid, water, silanes, ammonia borane, isopropanol and others. Rhodium[128] and palladium[129] based catalytic systems showed good reactivity and selectivity towards the formation of *(Z)*-alkenes with the application of molecular hydrogen as a reducing source. Due to low availability, price and toxicity of these metals, the replacement of the precious metals by the first-row, earth-abundant catalysts is highly desired.[37] Base metal-catalysed homogeneous hydrogenation of alkynes to *(Z)*-alkenes using hydrogen as reducing agent has hardly been investigated. The earliest example of a base-metal catalysed semireduction of alkynes to *(Z)*-alkenes was reported by Ugo`s group.[130] Phosphine cobalt carbonyl complexes were used for the selective reduction of 2-pentyne resulting in *(Z)*-2-pentene in a good yield. In 2017, Zhang and co-workers[131] successfully applied a cobalt complex in-situ formed from Co(OAc)$_2$·(H$_2$O)$_4$, NaBH$_4$ and ethylenediamine in proportion 1:2:8, respectively, for the chemoselective hydrogenation of C-C bonds. Iron catalysed hydrogenation of diphenylacetylene was mentioned in a work of Chirik et al.,[132] where the application of iron(0) dinitrogen complex initially resulted in the formation of *(Z)*-stilbene which was simultaneously converted to dibenzyl. The reaction is also known to be catalysed by Cr,[133] V,[134] Nb[135] and Cu[136] salts.

The high availability of manganese as the third most abundant transition metal in the earth's crust attracted its application in base-metal catalysis.[83–88] Inspired from the results on the highly selective transfer semihydrogenation of alkynes to *(Z)*-alkenes using a [Mn(II)-PNP][Cl$_2$] complex and ammonia borane as the hydrogen source.[137] I decided to develop Mn-catalysts able to use molecular hydrogen. Mn-catalysed direct hydrogenation of alkynes would be an advantageous process due to the fact that ammonia borane is a rather expensive and waste producing reducing agent. The investigation started by synthesizing a new, air and moisture stable ^{Ph}PNS-Mn pincer complex **Mn-8** by using a bench stable ^{Ph}PNS ligand **L3**. **Mn-8** can be readily synthesized by treatment of **L3** with 1 equiv. of Mn(CO)$_5$Br metal precursor in toluene at 100 °C for 16 h. The bright yellow complex was isolated in 85% yield and was characterized by NMR, IR, and mass spectrometry (Scheme 30).

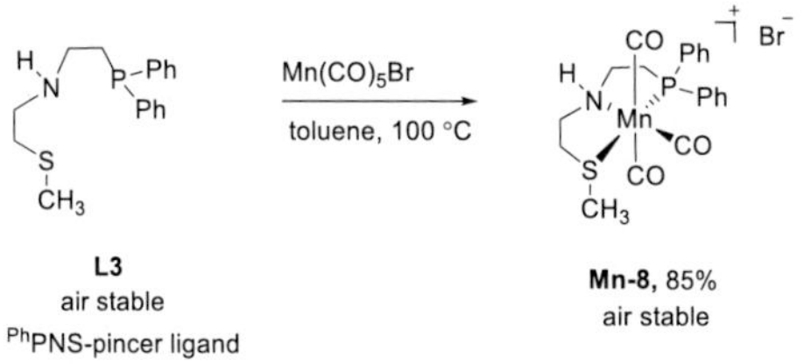

Scheme 30. Synthesis of **Mn-8**.

2.3.2 Optimisation of reaction conditions

For preliminary studies six manganese complexes were selected, among them a newly synthesized **Mn-8**. Initial attempts to hydrogenate diphenylacetylene proceeded using 1 mol% of manganese catalyst and 2.5 mol% of KO^{t}Bu in toluene at 60 °C applying 30 bar of H$_2$ for 16 h. The results for screening of the reaction conditions are summarised in the table below. The use of catalysts **Mn-1**, **Mn-2** and **Mn-3** provided unsatisfactory results (Table 5, entries 1-3). The application of **Mn-6** resulted in higher reactivity giving 48% conversion of starting material and 95:05 selectivity of **12a:13a** (Table 5, entry 4). The application of newly synthesized catalyst **Mn-8** resulted in full consumption of diphenylacetylene producing desired *(Z)*-stilbene in 88% GC yield, 1% of *(E)*-stilbene and 11% of dibenzyl as a result of over-hydrogenation (Table 5, entry 5). The catalyst **Mn-9**[38] showed better reactivity compared to **Mn-6** resulting in 60% conversion and a selectivity of 98:2 while dibenzyl was not detected (Table 5, entry 6). The control experiment showed that the reaction does not take place without the catalyst (Table 5, entry 7). Surprisingly, the use of K$_2$CO$_3$ or Cs$_2$CO$_3$ for the activation of the catalyst was not successful (Table 5, entries 8 and 9). Use of polar aprotic THF as a solvent resulted in 11% conversion of **11a** (Table 5, entry 10) while no reaction occurred when polar protic MeOH was used (Table 5, entry 11). Decreasing the hydrogen pressure to 20 bar helped to reduce formation of undesired overhydrogenation product (Table 5, entry 12). Additionally, no impact was observed when a drop of mercury was added to the reaction mixture, which suggests the homogeneous nature of the catalyst at these reaction conditions (Table 5, entry 13).[138]

Table 5. Optimisation of the reaction conditions.[a]

entry	[Mn] (1 mol%)	base (2.5 mol%)	conv. (%)[b]	ratio 12a:13a:14a (%)[b]
1	Mn-1	KO^{t}Bu	05	82:18:00
2	Mn-2	KO^{t}Bu	nr	nd
3	Mn-3	KO^{t}Bu	nr	nd
4	Mn-6	KO^{t}Bu	48	95:05:00
5	Mn-8	KO^{t}Bu	>99	88:01:11
6	Mn-9	KO^{t}Bu	60	98:02:00
7	-	KO^{t}Bu	nr	nd
8	Mn-8	K$_2$CO$_3$	nr	nd
9	Mn-8	Cs$_2$CO$_3$	17	90:10:00
10[c]	Mn-8	KO^{t}Bu	11	51:49:00
11[d]	Mn-8	KO^{t}Bu	nr	nd
12[e]	Mn-8	KO^{t}Bu	>99	96:01:03
13[e,f]	Mn-8+Hg	KO^{t}Bu	>99	93:01:06

[a] Reaction conditions: **11a** (1 mmol), 1 mol% of [**Mn**], base (2.5 mol%) in toluene (0.5 M) at 60 °C under 30 bar of H$_2$ for 16 h. [b] Determined by the GC analysis using *m*-xylene as internal standard. [c] Reaction in THF. [d] Reaction in methanol. [e] 20 bar of H$_2$. [f] One drop of mercury was added.

2.3.3 Scope of substrates

With optimised reaction conditions in hand a substrate scope for the selective semihydrogenation of alkynes was explored using a new Mn-PNS catalyst. A range of substrates bearing different electronic and steric properties were well tolerated and provided the corresponding *(Z)*-alkenes in good isolated yields with excellent chemoselectivity (Table 6). It should be noted that the substrates bearing electron withdrawing substituents were significantly more reactive than the ones bearing electron donating groups. Additionally, the hydrogenation of **11i**, bearing ester functionality, proceeded chemoselectively towards alkyne hydrogenation and the ester group remained intact. Importantly, alkynes which contain heterocycles (**11o-11q, 11x, 11y**) could also be applied and provided excellent reactivity and selectivity. Remarkably, no protodehalogenation of C-Cl and C-Br bond took place when 1-chloro-4-(phenylethynyl)benzene (**11l**) and 1-bromo-3-(phenylethynyl)benzene (**11n**) were applied as substrates. Moreover, my protocol was suitable for the application of triisopropyl(phenylethynyl)silane **11r** in the hydrogenation reaction. Due to higher steric hindrance of the substrate the reaction required 5 mol% of the **Mn-8**, a slightly higher hydrogen pressure of 30 bar and 24 hours of the reaction time, resulting in 76% isolated yield of *(Z)*-triisopropyl(styryl)silane as a single isomer without formation of alkane as a by-product. Furthermore, the reduction of aryl-alkyl alkynes, including protected propargylic alcohols **11s−11y** led to the formation of corresponding *(Z)*-allylic alcohols demonstrating the wide scope of substrates.

Table 6. (*Z*)-Selective hydrogenation of alkynes catalysed by **Mn-8**.[a]

Alkene	Cat. [mol%]	t [h]	ratio 12:14[b]	Y of 12[e]	Alkene	Cat. [mol%]	t [h]	ratio 12:14[b]	Y of 12[e]	Alkene	Cat. [mol%]	t [h]	ratio 12:14[b]	Y of 12[e]
12b	2	12	95:05	81%	12j	3	16	93:07	86%	12r[c]	5	24	100:0	76%
12c[e]	2	16	94:06	79%	12k[c]	3	16	92:08	78%	12s[c]	3	16	90:10	69%
12d	2	16	96:04	94%	12l	1	12	92:08	85%	12t	2	16	100:0	91%
12e	2	16	93:07	83%	12m	1	16	90:10	82%	12u	1	16	89:11	81%
12f[e]	1	16	94:06	90%	12n[d]	1	16	89:11	75%	12v	2	16	100:0	94%
12g[d]	1	16	93:07	80%	12o	1	8	92:08	77%	12w	1	16	90:10	73%
12h[c]	2	20	91:09	72%	12p[c]	1	16	93:07	90%	12x	1	16	95:05	76%
12i[e]	2	16	91:09	82%	12q[c]	2	16	95:05	79%	12y	1	12	94:06	82%

[a] Reaction conditions: alkyne (0.5 mmol), **Mn-8**, KO*t*Bu 2.5 equiv. to the **Mn-8** in toluene (0.5 M) at 60 °C under 20 bar of H$_2$. [b] Determined by the NMR analysis using CH$_2$Br$_2$ as internal standard. [c] 30 bar of H$_2$. [d] 50 °C. [e] Isolated yield.

Attempts to hydrogenate phenylacetylene were unsuccessful revealing that the developed catalytic system was incompatible with terminal alkynes. Additionally, substrates with carbonyl or nitrile group were not tolerated as they were also reduced under applied reaction conditions. Moreover alkyl-alkyl substituted substrates were inactive in the developed catalytic system (Scheme 31).

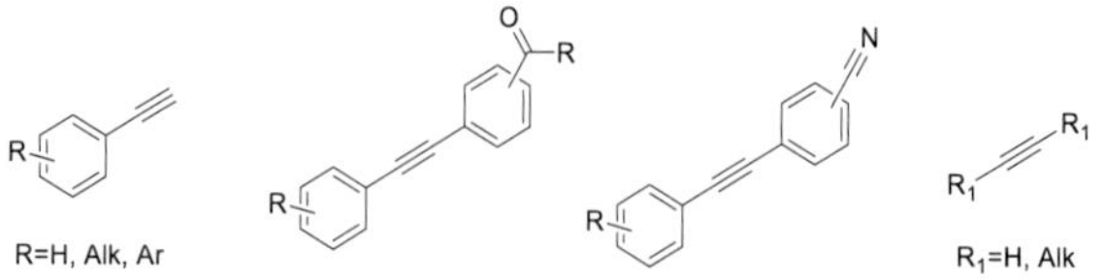

Scheme 31. Scope limitations.

Furthermore, a gram-scale synthesis of *(Z)*-stilbene could also be achieved using only 0.5 mol% of **Mn-8**, leading to the formation of 99% yield of the desired product (Scheme 32) implying that described protocol could be suitable for industrial production of *(Z)*-alkenes.

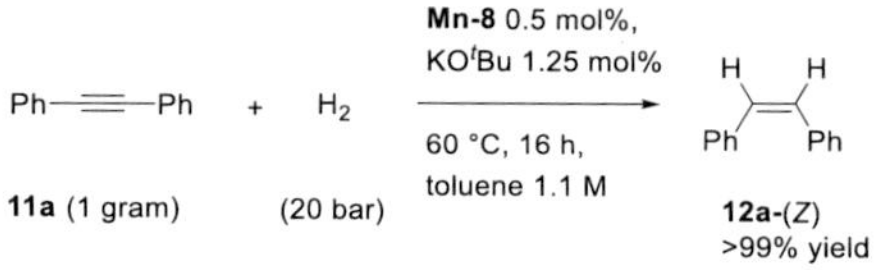

Scheme 32. Gram-scale synthesis of *(Z)*-stilbene.

2.3.4 Mechanistic studies

To prove whether the described reaction proceeds *via* metal-ligand cooperativity I attempted to synthesize the corresponding manganese **N-Me** derivative of **Mn-8**. Unfortunately, the formation of the **Mn-8(N-Me)** catalyst was not successful after several attempts, using different solvents and temperatures. However, it was possible to prepare corresponding **N-Me** manganese complex for **Mn-6**, which also showed reactivity in the hydrogenation. As expected, the methylated complex **Mn-6(N-Me)** appeared to be inactive in the hydrogenation of diphenylacetylene under optimised reaction conditions, indicating that the formation of the N-H is critical for the activity of the catalyst (Scheme 33a). To exclude the possibility of the interaction between the substrate and the catalyst, diphenylacetylene was stoichiometrically added to the **Mn-8*** (active species), *in-situ* formed by the addition of KO^{t}Bu to the **Mn-8** complex and heated for 24 hours in C$_6$D$_6$ (Scheme 33b). The chemical shift of the activated **Mn-8*** remained unaffected, indicating that no coordination of the alkyne and potentially inhibition occurs.

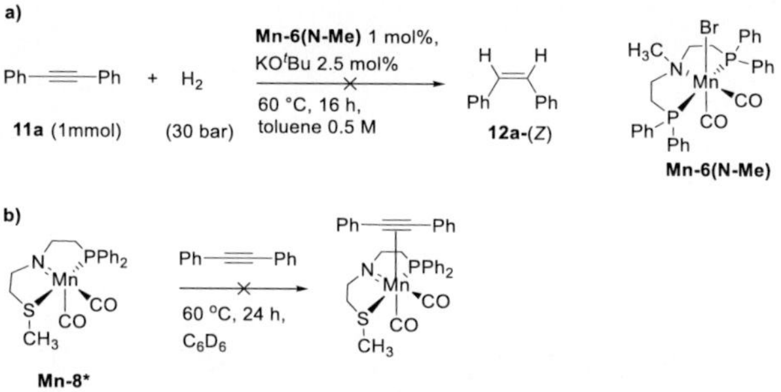

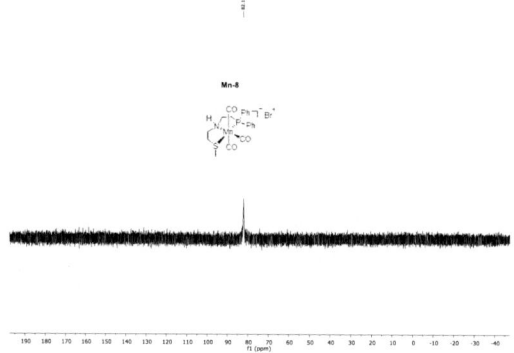

Scheme 33. Mechanistic studies to prove metal-ligand cooperativity.

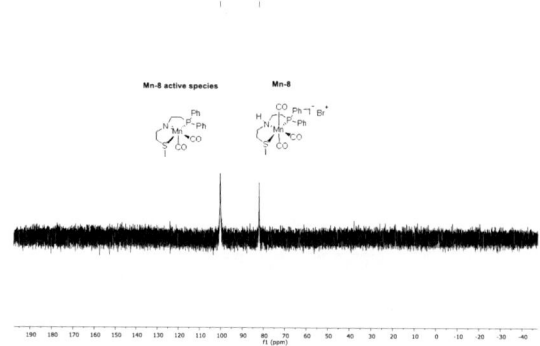

Figure 9. Mn-8 complex.

Figure 10. Reaction between **Mn-8** and KO^{t}Bu (16 h, 60 °C).

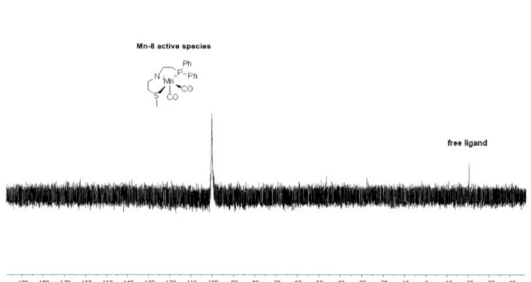

Figure 11. Reaction mixture (with diphenylacetylene) after 5 h at 60 °C.

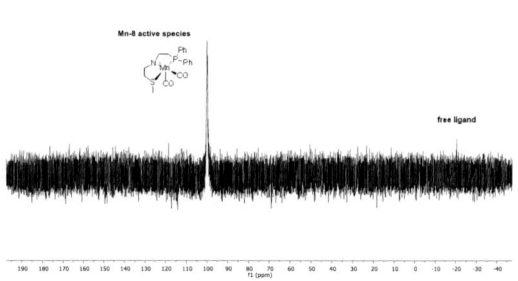

Figure 12. Reaction mixture (with diphenylacetylene) after 24 h at 60 °C.

2.3.5 Proposed reaction mechanism

Based on observed experimental results I propose that the reaction proceeds *via* metal-ligand cooperativity following an outer-sphere pathway (Scheme 34).[117–119] Thus, the catalytic cycle begins with the addition of the molecular hydrogen to the metal site of the catalyst and formation of the intermediate **A**. Next, heterolytic cleavage of the H-H bond takes place leading to the hydrogenated catalyst **B** *via* transition state **TS-1**. A proton and a hydride are transferred simultaneously from the intermediate **B** to the substrate giving an intermediate **C**, which later releases the desired product **12a** and the active catalyst **Mn-8***.

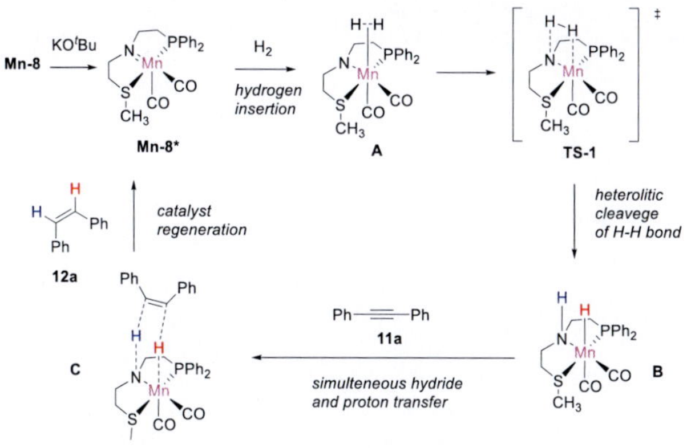

Scheme 34. Proposed reaction mechanism of alkyne hydrogenation using **Mn-8** complex.

2.3.6 Summary

In conclusion, a new manganese-catalysed semihydrogenation of alkynes using molecular hydrogen as reducing agent has been developed. The reaction proceeds under mild reaction conditions and provides the desired *(Z)*-alkenes with very high selectivity. The applied catalyst **Mn-8** can be synthesised from a commercially available manganese precursor and an air stable and readily available [Ph]PNS-pincer ligand highlighting the practicability of the developed protocol. The **Mn-8** catalyst shows good reactivity and chemoselectivity and tolerates a variety of functional groups and heterocycles leading to practical synthesis of *(Z)*-olefins as well as allylic alcohols.

2.4 Hydrogenation and dehydrogenation of heterocycles with the application of manganese pincer complexes

2.4.1 Introduction

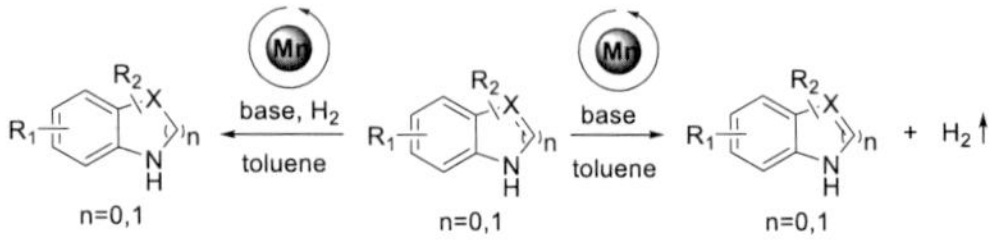

Scheme 35. Manganese-catalysed (de)hydrogenation of heterocycles.

Transition metal catalysed hydrogenation of polar bonds is a well-accepted and widely used method for the synthesis of a diverse set of value-added products such as alcohols, amines, saturated heterocycles etc.[124] However, most of the reports focus on using rare and expensive transition metals or heterogeneous catalysts, which may require harsh reaction conditions resulting in a low functional group tolerance. The replacement of noble-metals by sustainable base-metals is currently getting increased attention due to their lower toxicity and ubiquitous abundance.[37] On the other hand, saturated and unsaturated heterocycles are considered as liquid organic hydrogen carriers (LOHC) due to their reversible dehydrogenating properties. Using N-containing heterocycles as LOHC allows avoiding problems associated with commonly studied LOHC reagents including ammonia borane, sodium borohydride, metal hydrides. First of all, they are abundant and economically advantageous. Second, the dehydrogenation process for these molecules is endothermic, which prevents uncontrolled thermal reactions. Thus, N-containing heterocycles are considered to be a good alternative if compared to hydrocarbons due to the lower energy barrier for de/hydrogenation processes.[139] Examples of a single catalyst which are able to catalyse both, the hydrogenation and dehydrogenation process are very rare in the literature. Fujita and co-workers studied iridium complexes for this transformation.[140] Later, Crabtree and co-workers[141] and Albrecht and co-workers[142] reported the use of iridium complexes for the catalytic hydrogenation and dehydrogenation of N-heterocycles in water. In addition, Fischmeister and co-workers reported a mild reversible hydrogenation of quinoline derivatives using an iridium-based catalyst.[143] Although, the field is predominant by the application of iridium catalysts, Jones and co-workers focused on using base-metals such as Fe[20] and Co[144] for this transformation. However, Mn-based systems still remain unknown. Therefore, the development of single catalysts for the reversible dehydrogenation process is interesting and desired (Scheme 35). Recently an increasing number of reports featuring the high reactivity of Mn-complexes for the hydrogenation of organic molecules were reported.[83–88] To the best of my knowledge, only few reports addressing the reduction of heteroaromatic systems were published.[145] Based on my interest in manganese catalysis as well as hydrogenation and dehydrogenation reactions I decided to explore the hydrogenation of indoles as representatives of N-heterocyclic compounds. The indole scaffold is considered to be one of the most important organic frameworks for the discovery of new drugs as many of the indole derived compounds play a significant role in nature. Among them are tryptophan, an α-amino acid which is essential to humans, the neurotransmitter serotonin, and melatonin, a hormone which regulates the sleep-awake cycles. The indoline skeleton is equally important and it is found in numerous bioactive compounds, pharmaceuticals, herbicides, and insecticides.[146] Hydrogenation of indoles is a difficult task due to the high stability of the aromatic heterocyclic ring. Among the conventional methods to achieve saturated heterocycles we may highlight the use of $NaBH_3CN$. It is one of the most used methods, however due to the use of superstoichiometric amounts of the hydride source and the generation of high amounts of waste, such as cyanides, other improved systems are still desired. The catalytic hydrogenation using hydrogen gas as the reducing agent is an attractive process due to the low cost of hydrogen, atom economy and minimal waste generation.

2.4.2 Optimisation of reaction conditions

Encouraged by the previous results of our group I was interested to find out whether a bench stable Mn-PNP **Mn-6** catalyst would be active enough for the hydrogenation of *N*-containing heterocyclic compounds. Simple unsubstituted indole **15a** was chosen as a model substrate to test the above-mentioned reaction. Upon running the reaction for 24 h at 100 °C, 50 bar of H_2, with 2 mol% of the catalyst and 5 mol% of base indoline **16a** was formed in high yield (85%) (Table 7, entry 1). The yield did not increase when the reaction was performed in polar protic *t*-amyl alcohol and polar aprotic dioxane as solvent (Table 7, entries 2-3). Increasing the reaction time to 36 h led to indoline **16a** in 92% yield (Table 7, entry 4). Using $CsOH*H_2O$ instead of KO^tBu provided the product with the same yield (Table 7, entry 5). The application of other bases did not improve the reaction outcome (Table 7, entries 6-8). Decreasing the catalyst loading to 1 mol% resulted in 47% yield of indoline only (Table 7, entry 9). The application of $Mn(CO)_5Br$ precursor in the reaction resulted in the full recovery of the indole (Table 7, entry 10), which highlights the crucial role of the ligand.

Table 7. Optimization of the reaction conditions.[a]

entry	Mn-6 (mol%)	base (mol%)	solvent	yield (%)[b]
1[c]	Mn-6 (2)	KO^{t}Bu (5)	toluene	85
2[c]	Mn-6 (2)	KO^{t}Bu (5)	TAA	57
3[c]	Mn-6 (2)	KO^{t}Bu (5)	dioxane	18
4	Mn-6 (2)	KO^{t}Bu (5)	toluene	92
5	Mn-6 (2)	CsOH*H$_2$O (5)	toluene	92
6	Mn-6 (2)	Cs$_2$CO$_3$ (5)	toluene	88
7	Mn-6 (2)	K$_3$PO$_4$ (5)	toluene	36
8	Mn-6 (2)	NaO^{t}Bu (5)	toluene	71
9	Mn-6 (1)	KO^{t}Bu (2.5)	toluene	47
10	Mn(CO)$_5$Br (2)	KO^{t}Bu (5)	toluene	nd

[a] Reaction conditions: **15a** (0.25 mmol), [Mn], base in selected solvent (0.25 M) at 100 °C under 50 bar of H$_2$ for 36 h. [b] Determined by GC analysis using dodecane as internal standard. [c] Reaction time is 24 h. TAA = t-amyl alcohol.

2.4.3 Scope of substrates

In order to demonstrate the potential and applicability of the newly developed catalytic hydrogenation system, a range of substituted indoles **15a-15o** were tested under the optimised reaction conditions (Table 8). Different substituted indoles were efficiently and selectively hydrogenated to the corresponding indolines with good to very good yields. Simple substituted indoles with methyl groups in the C-6 and C-7 positions were also well tolerated and led to the desired products **16b** and **16c** in very good yields. An elevated temperature was needed when 5-methoxyindole was applied in the reaction, resulting in 85% of the corresponding indoline **16d**. Remarkably, methyl indole-5-carboxylate was selectively hydrogenated, yielding the desired indoline in 97% yield, with the ester group remaining intact. Halogen containing substrates were also tolerated, resulting in high yields for the products **16f-16i**. It is worth mentioning that no hydrodehalogenation occurred under the optimised reaction conditions. Scale up experiments were performed resulting in high yields of the corresponding indolines. Quantitative yields were observed for the substrates **15e**, **15f** and **15h** which indicates the high potential of this transformation. In addition, the hydrogenation of C3-substituted indoles could be performed. The application of an elevated reaction temperature (130 °C) and a higher catalyst loading (5 mol%) were required for the reaction to proceed successfully.

Table 8. Manganese-catalysed hydrogenation of indoles.[a]

Reaction scheme: **15** + H₂ (50 bar) → **16**, conditions: Mn-6, KO^{t}Bu, 100 °C, 36 h, toluene 0.25 M.

Products:
- **16a**, 83%
- **16b**, 91%[b]
- **16c**, 69%
- **16d**, 85%[b,c]
- **16e**, 97%[b]
- **16f**, 99%[b]
- **16g**, 86%
- **16h**, 99%[b]
- **16i**, 86%[b]
- **16j**, 78%[c,d]
- **16k**, 78%[c,d]
- **16l**, 65%[c,d]
- **16m**, 72%[c,d]
- **16n**, 76%[c,d]
- **16o**, 69%[c,d]

[a] Reaction conditions: **15** (0.25 mmol), **Mn-6** (2 mol%) and KO^{t}Bu (5 mol%) in toluene (0.25 M) at 100 °C under 50 bar of H₂ for 36 h. Isolated yields provided. [b] Yields for scale-up experiment (2.5 mmol of starting material used). [c] 130 °C. [d] **Mn-6** (5 mol%) and KO^{t}Bu (12.5 mol%).

In addition to indoles other different N-containing heterocycles could be applied in the transformation (Table 9). Unsubstituted and substituted quinoxalines **17a-c** as well as benzoxazine **17d** were successfully applied in the reaction resulting in quantitative yields of the corresponding products.

Table 9. Manganese-catalysed hydrogenation of N-containing heterocycles.[a]

Reaction scheme: **17** (X=N, O) + H₂ (50 bar) → **18**, conditions: Mn-6, KO^{t}Bu, 120 °C, 24 h, toluene 0.25 M.

Products:
- **18a**[b,c], >99%
- **18b**, >99%
- **18c**[b,d], >99%
- **18d**, >99%

[a] Reaction conditions: **17** (0.25 mmol), **Mn-6** (1 mol%) and KO^{t}Bu (2.5 mol%) in toluene (0.25 M) at 120 °C under 50 bar of H₂ for 24 h. Isolated yields provided. [b] 140 °C. [c] **Mn-6** (2 mol%) and KO^{t}Bu (5 mol%). [d] **Mn-6** (3 mol%) and KO^{t}Bu (7.5 mol%).

Considering the above results, I decided to investigate whether a manganese-catalysed dehydrogenation would also be accessible demonstrating for the first time the applicability of manganese complexes in both, hydrogenation and dehydrogenation of amines. Thus, **Mn-6** was applied to catalyse the dehydrogenation of N-containing heterocycles under oxidant-free conditions. By using only 1 mol% of the catalyst at 120 °C for 24 h indoline was successfully dehydrogenated leading to 89% of indole. The liberated hydrogen gas was detected using GC. Although, the hydrogen storage capacity for indoline is relatively low, 1.7 wt% compared to the current favorite, the N-ethylcarbazole/dodecahydro-N-ethylcarbazole system with a capacity of storing 5.8 wt% of hydrogen, its conversion to indole requires low catalyst loading and comparable low temperatures which makes the developed catalytic system potentially interesting for the LOHC (liquid organic hydrogen carrier) concept. A higher catalyst loading (5 mol%) and temperature (150 °C) were required for the successful dehydrogenation of 3-methylindoline leading to product **15j** in 92% isolated yield. Harsher conditions had to be applied for the dehydrogenation of 1,2,3,4-tetrahydroquinoxaline as well as substituted 1,2,3,4-tetrahydroquinoxalines. Thus, by using 5 mol% of the catalyst and conducting the reaction at 160 °C for 24 h 95% of quinoxaline could achieved. Substituted 1,2,3,4-tetrahydroquinoxalines underwent dehydrogenation by using 10 mol% of the catalyst at 160 °C for 24 h (Table 10).

Table 10. Manganese-catalysed dehydrogenation of N-heterocycles.[a]

		R:	H	Me	R:	H	Me	Ph
			15a, 89%	**15j**, 92%		**17a**, 95%	**17b**, 73%	**17c**, 97%
Mn-6			1 mol%	5 mol%		5 mol%	10 mol%	10 mol%
T, °C			120 °C	150 °C		160 °C	160 °C	160 °C

[a] Reaction conditions: **16** or **18** (0.25 mmol), **Mn-6** (1-10 mol%) and KO^{t}Bu (2.5-25 mol%) in toluene (0.25 M) at 120-160 °C for 24 h. Isolated yields provided.

2.4.4 Mechanistic studies and proposed reaction mechanism

The proposed catalytic cycle for the hydrogenation of indole and dehydrogenation of indoline is depicted in Scheme 36. The first step for the hydrogenation process is the activation of **Mn-1** using the base (KO^{t}Bu) and formation of an active species **A**. Next, species **A** reacts with hydrogen forming H-N-Mn-H species **B**, which successfully hydrogenates indole to indoline. The dehydrogenation reaction begins also with the formation of an active species **A** which dehydrogenates indoline to indole and becomes species **B** which releases a hydrogen molecule afterwards.

In order to support the described proposed mechanism and to understand whether this reaction proceeds *via* a metal-ligand cooperative mechanism,[117,118] the hydrogenation of indole **15a** was performed using N-Me substituted **Mn-6** catalyst (Scheme 37). As expected, no formation of indoline was observed under the applied reaction conditions, indicating the crucial role of species **B** for the catalytic cycle.

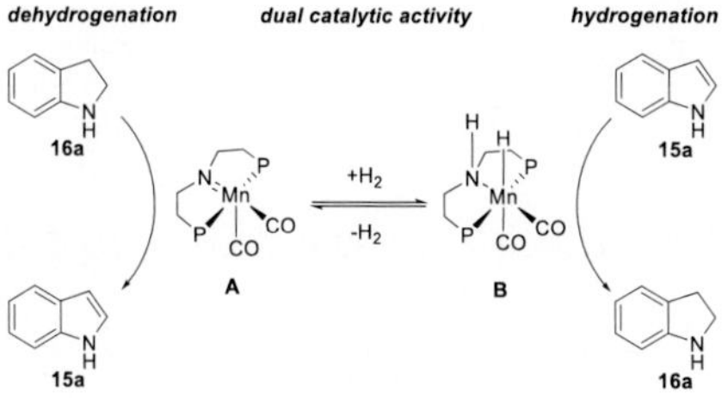

Scheme 36. Proposed reaction mechanism of reversible indole hydrogenation and dehydrogenation using **Mn-6** complex.

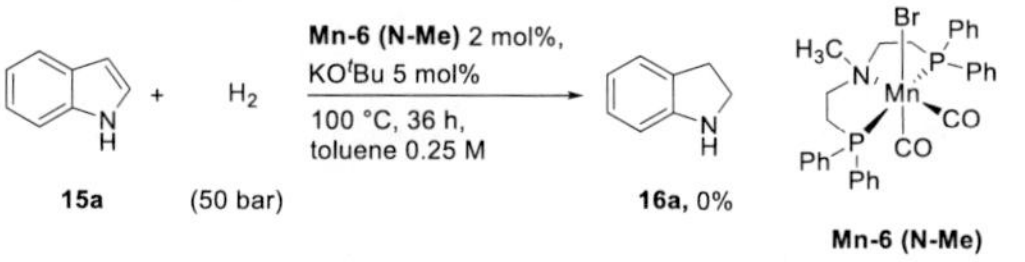

Scheme 37. Mechanistic studies.

2.4.5 Summary

In conclusion, for the first time a single manganese catalyst that can catalyse both, the hydrogenation and acceptorless dehydrogenation of *N*-containing heterocycles is presented. The products of both hydrogenation and dehydrogenation reactions can be isolated in good to excellent yields and high chemoselectivity. The applied catalyst **Mn-6** is scalable, air and moisture stable and can be synthesized using a readily available manganese precursor and ^{Ph}PNP-pincer ligand, highlighting the practicability of the developed protocol. Mechanistic studies indicate a metal-ligand cooperative catalysis.

2.5 Manganese-catalysed hydrogenation of nitroarenes

2.5.1 Introduction

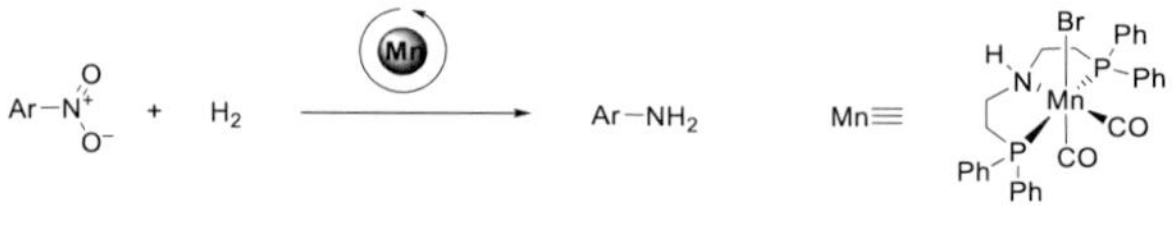

Scheme 38. Manganese-catalysed hydrogenation of nitroarenes.

The reduction of nitroarenes to anilines represents one of the most essential reactions in organic chemistry. Variety of procedures were developed to achieve anilines by hydrogenation of nitroarenes.[147–149] This straightforward approach distinguishes minimum waste generation as it gives water as a sole by-product.[150] Hydrogenation of nitroarenes has also a great importance in industry due to the high demand of anilines for pharmaceuticals, dyes and agrochemical production, as well as, polyurethanes synthesis. One of the commonly used reaction transforming nitroarenes into anilines is the Bechamp reduction.[151] Despite of a high functional group tolerance, that process exhibits considerable drawbacks. Bechamp's procedure requires the use of corrosive hydrochloric acid and superstochiometric amounts of iron or iron salts leading to significant amounts of waste. Due to the high importance of substituted anilines, more economically beneficial methods are required. At present, the most commonly applied procedure is catalytic hydrogenation of nitroarenes utilizing toxic and expensive Pd/C or pyrophoric Ni-Rainey, which often suffer from low chemoselectivity. The desired chemoselectivity can be achieved by the modification of the standard catalysts. The application of modifiers usually decreases the reactivity as a result of coverage of the active site and also involves laborious and complex preparation.[147]

Major part of the reports for the catalytic hydrogenation of nitroarenes is focused on the development and modification of heterogeneous catalysts.[147–149] However, very few works were devoted to homogeneous reduction of nitroarenes. Clearly, it is easier to modify homogeneous metal catalysts by the application of different ligands or the metal itself to achieve high selectivity. Currently it is hard to compete with high performances of heterogeneous catalysts, nevertheless simple molecular catalysts can be helpful for the synthesis of specific pharmaceuticals, which require high selectivity and low toxicity of the catalyst. For this reason range of protocols were developed using homogeneous catalysts based on the noble metals including Au,[152] Ir,[153] Pd,[152,154,155] Pt,[154] Rh,[153,156] Ru.[154,157,158] Therefore, the replacement of noble-metal catalysts by earth-abundant alternatives is highly desirable in the context of a sustainable chemistry. The first attempts to apply base metal catalysis to hydrogenate nitroarenes to anilines were performed by J. F. Knifton in 1976.[158] Fe(CO)$_3$(PPh$_3$)$_2$ and Fe(CO)$_3$(AsPh$_3$)$_2$ were applied in low catalytic loading leading to the selective formation of aniline under moderately mild conditions. Regardless of these impressive results only few other reports were conducted on this topic. In 2004, Chaudhari *et al.* performed hydrogenation of nitroarenes in aqueous/organic biphasic medium using Fe/EDTANa$_2$ system.[159] Presence of biphasic system allows better separation of the product from the catalyst but also slows down the reaction. Good chemoselectivities were observed despite using relatively high reaction temperature of 150 °C. In 2013, Beller and co-workers reported iron-based complex for the catalytic hydrogenation of nitroarenes.[160] The developed system operates under relatively mild reaction conditions and tolerates various functional groups. However, up to 2 equiv. of a strong acid, such as trifluoroacetic acid, had to be added for a significant catalyst activity. Lack of reports regarding application of base-metals for hydrogenation of nitroarenes was my motivation to employ the **Mn-6** catalyst, which can activate molecular hydrogen as powerful, inexpensive and environmentally friendly reducing agent. Recently, the application of manganese, as the third most abundant metal in the Earth`s crust, for the catalytic hydrogenation of organic molecules considerably increased.[83–88] To the best of my knowledge, no manganese-catalysed reduction of nitroarenes to anilines have been previously reported (Scheme 38). [147–149]

2.5.2 Optimisation of reaction conditions

The investigation started with the application of the catalyst **Mn-6** which can be easily prepared from commercially available ligand and metal precursor. Nitrobenzene was chosen as a model substrate in order to achieve optimised reaction conditions. First attempt to hydrogenate nitrobenzene proceeded using 5 mol% of **Mn-6** and 12.5 mol% of KO^{t}Bu in toluene at 130 °C applying 50 bar of H$_2$ for 24 h. **Mn-6** proved to be active towards the reduction of nitrobenzene producing desired aniline in 59% GC yield (Table 11, entry 1). In the next step other solvents were tested. The use of polar and aprotic 1,4-dioxane resulted in 44% yield of **20a** (Table 11, entry 2) while the application of t-amyl alcohol led to 21% yield (Table 11, entry 3). Hence toluene was established as the best solvent for this transformation, different bases were investigated. When Cs$_2$CO$_3$ was used for the activation of the catalyst, the yield dropped to 35%, while CsOH*H$_2$O showed similar reactivity to KO^{t}Bu affording **20a** in 52% yield (Table 11, entries 4 and 5). Delightfully, the use of cheap and highly available K$_2$CO$_3$ in the reaction led to the formation of aniline in 87% yield (Table 11, entry 6). Lastly, KH was tested leading to unsatisfactory result with 27% yield of the desired product (Table 11, entry 7). Next, reaction temperature was increased to 140 °C and pressure of H$_2$ to 80 bar which allowed us to reach over 99% yield of the aniline in both cases (Table 11, entries 8 and 9). For the next studies I chose to keep the increased pressure. As expected, without the addition of the catalyst or the base the reaction did not take place (Table 11, entries 10 and 11).

Table 11. Optimisation of the reaction conditions.[a]

№	base (12.5 mol%)	GC yield. (%)[b]
1	KO^{t}Bu	59
2[c]	KO^{t}Bu	44
3[d]	KO^{t}Bu	21
4	Cs$_2$CO$_3$	35
5	CsOH*H$_2$O	52
6	K$_2$CO$_3$	87
7	KH	27
8[e]	K$_2$CO$_3$	>99
9[f]	K$_2$CO$_3$	>99
10[f]	-	<5
11[f,g]	K$_2$CO$_3$	nr

[a] Reaction conditions: nitrobenzene (0.25 mmol), **Mn-6** (5 mol%), base (12.5 mol%) in toluene (0.25 M) at 130 °C under 50 bar of H$_2$ for 24 h. [b] Determined by the GC analysis using dodecane as internal standard. [c] 1,4-Dioxane is used as a solvent. [d] t-Amyl alcohol is used as a solvent. [e] 140 ºC. [f] 80 bar of H$_2$. [g] No catalyst.

2.5.3 Scope of substrates

Having optimised reaction conditions substrate scope for the hydrogenation of nitroarenes was explored. A range of alkyl substituted nitroarenes were well tolerated and provided the corresponding anilines in excellent isolated yields up to 97% (Table 12, **20b-20e**). It should be noted that the halogenated substrates were well tolerated affording high yields of the desired aniline derivatives (Table 12, **20f-20i**). Remarkably, no protodehalogenation of C-Hal bond took place and trifluoromethyl group in *meta* position of the aromatic ring is also tolerated (**20j**). The developed system is able to chemoselectively reduce nitro group in the presence of the double bond (**20k**), ester group (**20p, 20q**), amido functionality (**20r**) and compounds containing sulfonamide fragment (**20u**). Other functional groups such as ether, thioether as well as amino group were well tolerated and the desired anilines could be isolated with very high yields up to 99% (Table 12, **20l-20o** and **20s, 20t**). Finally, 1-nitronaphthalene (**19v**) was successfully applied providing naphthalen-1-amine (**20v**) in 75% yield.

Table 12. Selective hydrogenation of nitroarenes catalysed by **Mn-6**.[a]

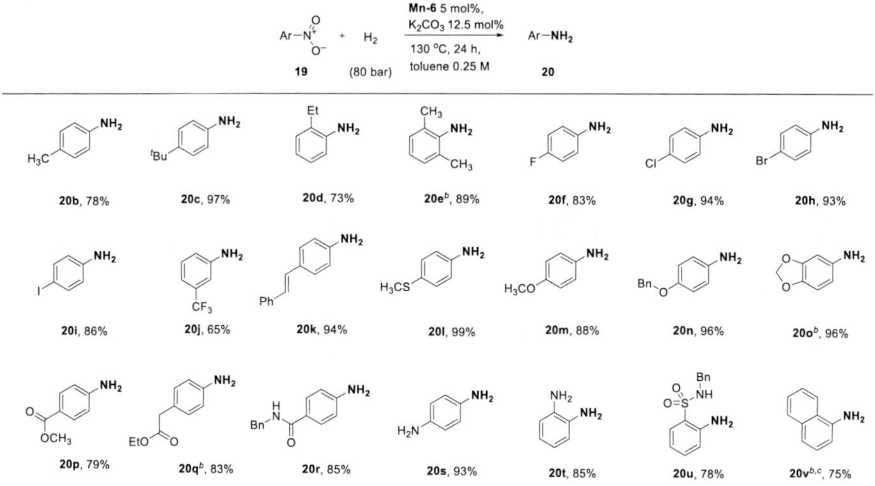

$$Ar-NO_2 \ (19) + H_2 \ (80 \ bar) \xrightarrow[130 \ °C, \ 24 \ h, \ toluene \ 0.25 \ M]{Mn-6 \ 5 \ mol\%, \ K_2CO_3 \ 12.5 \ mol\%} Ar-NH_2 \ (20)$$

Product	Yield
20b (4-Me-C₆H₄-NH₂)	78%
20c (4-tBu-C₆H₄-NH₂)	97%
20d (2-Et-C₆H₄-NH₂)	73%
20e[b] (2,3-diMe-C₆H₃-NH₂)	89%
20f (4-F-C₆H₄-NH₂)	83%
20g (4-Cl-C₆H₄-NH₂)	94%
20h (4-Br-C₆H₄-NH₂)	93%
20i (4-I-C₆H₄-NH₂)	86%
20j (3-CF₃-C₆H₄-NH₂)	65%
20k (4-styryl-NH₂)	94%
20l (4-SMe-C₆H₄-NH₂)	99%
20m (4-OMe-C₆H₄-NH₂)	88%
20n (4-OBn-C₆H₄-NH₂)	96%
20o[b] (3,4-methylenedioxy-C₆H₃-NH₂)	96%
20p (4-OC(O)CH₃-C₆H₄-NH₂)	79%
20q[b] (4-CH₂CO₂Et-C₆H₄-NH₂)	83%
20r (4-NHC(O)Bn-C₆H₄-NH₂)	85%
20s (4-NH₂-C₆H₄-NH₂)	93%
20t (2-NH₂-C₆H₄-NH₂)	85%
20u (2-SO₂NHBn-C₆H₄-NH₂)	78%
20v[b,c] (1-naphthyl-NH₂)	75%

[a] Reaction conditions: nitroarene (0.25 mmol), **Mn-6** (5 mol%), K₂CO₃ (12.5 mol%) in toluene (0.25 M) at 130 °C under 80 bar of H₂, isolated yields are provided. [b] **Mn-6** (10 mol%), K₂CO₃ (25 mol%). [c] 48 h.

Attempts to hydrogenate alkyl substituted nitro compounds were unsuccessful. Application of nitrophenols and substrates with unsubstituted sulfamide group led to a decomposition of starting material. Double nitro substituted compounds were inactive possibly due to a coordination of the substrate to the catalyst. Additionally, substrates with carbonyl or nitrile group were not tolerated as they were also reduced under applied reaction conditions. Moreover, substrates with acid functionality as well as unsubstituted amide were inactive in the developed catalytic system. Heterocyclic compounds and compounds with halogens in *ortho* position were inactive as well (Scheme 39).

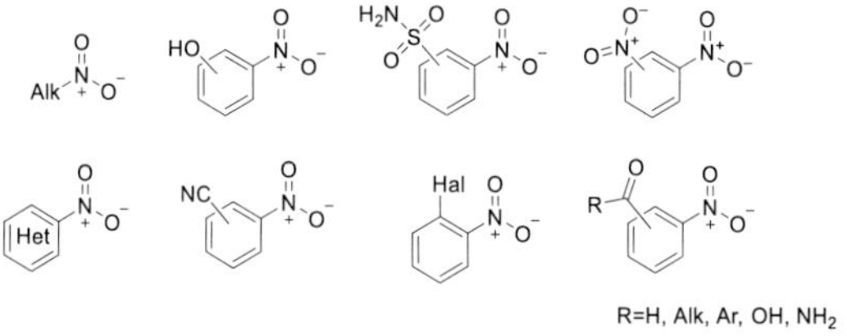

Scheme 39. Scope limitations.

Additionally, a gram-scale synthesis of 4-iodoaniline could also be performed using the optimised reaction conditions, remarkably, only 4 mol% of the catalyst were required for the full conversion. The product was formed in 78% yield (Scheme 40) implying feasibility of the described protocol.

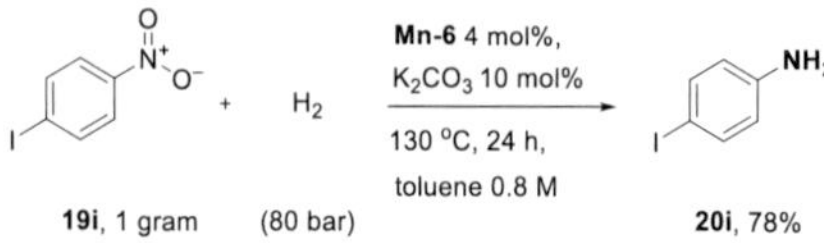

Scheme 40. Gram-scale synthesis of 4-iodoaniline.

In order to demonstrate a general applicability of the developed method I decided to perform the hydrogenation of the intermediate **19w** in the synthesis of vortioxetine, an antidepressant used to treat major depressive disorder. A newly developed synthetic route includes a nucleophilic substitution of 2,4-dimethylbenzenethiol with 1-chloro-2-nitrobenzene to afford the desired intermediate **19w**.[161] Under optimised reaction conditions the formed nitrophenylsulfane derivative (**19w**) undergoes catalytic hydrogenation leading to the desired thioaniline derivative (**20w**) in 74% yield (Scheme 41). The reaction of **20w** with 2-chloro-N-(2-chloroethyl)ethanamine hydrochloride provides the desired vortioxetine.

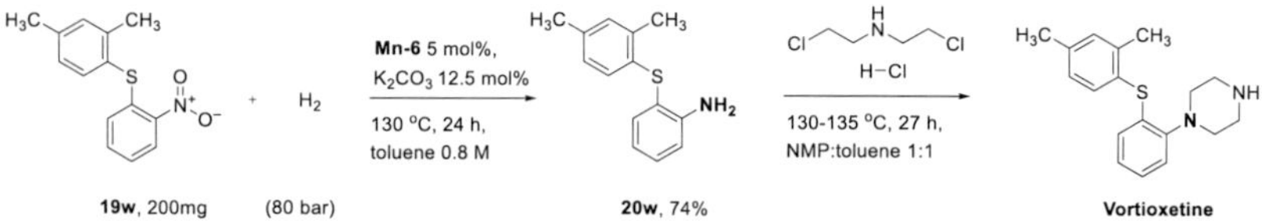

Scheme 41. Synthesis of vortioxetine intermediate.

2.5.4 Mechanistic studies

To prove whether the described reaction proceeds *via* metal-ligand cooperativity I performed the hydrogenation reaction using the corresponding manganese **N-Me** derivative of **Mn-6**. As expected, the methylated complex **Mn-6 (N-Me)** appeared to be inactive in the hydrogenation of nitrobenzene under optimised reaction conditions, indicating that the presence of the N-H is critical for the reaction to proceed (Scheme 43a). There are two commonly studied pathways for the hydrogenation of nitroarenes to anilines. The first one is a direct pathway where the reduction proceeds *via* the formation of nitrosoarene and hydroxylamine intermediates. The second one can occur when an azoxy compound is formed by condensation of nitrosoarene and hydroxylamine, which later undergoes reduction to azo- and hydrazo-compounds (Scheme 42). In order to investigate the reaction mechanism of the studied catalytic system, possible intermediates were submitted to the standard reaction conditions. N-phenylhydroxylamine (**21**, Scheme 43b), azobenzene (**22**, Scheme 43b) and 1,2-diphenylhydrazine (**23**, Scheme 43b) were tested. The reduction of N-phenylhydroxylamine led to the formation of 49% of aniline, when azobenzene and 1,2-

diphenylhydrazine provided only 7% and 10% of aniline respectively. The results suggest that hydrogenation of nitroarenes in the developed catalytic system rather undergoes direct hydrogenation. In the case of the formation of the unwanted azo and hydrazo compounds, **Mn-6** can partially transform them to the desired anilines. It should be noted that no accumulation of intermediates such as hydroxylamine or azo-, hydrazo- and azoxy-compounds was observed.

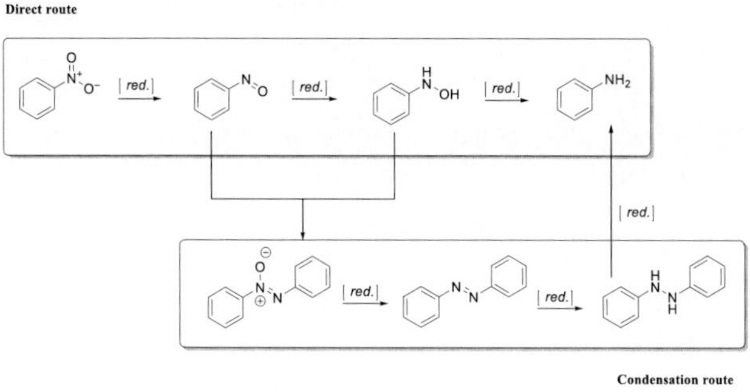

Scheme 42. Possible pathways for the hydrogenation of nitroarenes.

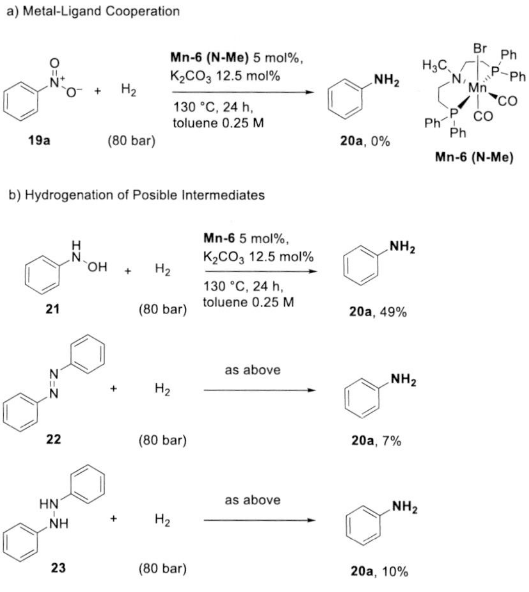

Scheme 43. Mechanistic studies.

2.5.5 Summary

In conclusion, manganese-catalysed hydrogenation of nitroarenes was developed using molecular hydrogen as reducing agent. The applied catalyst **Mn-6** can be synthesized from a commercially available manganese precursor as well as an air stable and readily available ^{Ph}PNP-pincer ligand highlighting the high feasibility of the developed protocol. The reaction proceeds under relatively mild reaction conditions and provides the desired aniline derivatives in excellent yields. The **Mn-6** catalyst shows very good reactivity and chemoselectivity and tolerates a variety of functional groups leading to practical synthesis of anilines. The performed mechanistic studies suggested that the reaction proceeds *via* metal-ligand cooperative catalysis *via* a direct pathway to achieve desired anilines.

2.6 Intermolecular alkylation of amines with alcohols to form heterocycles

2.6.1 Introduction

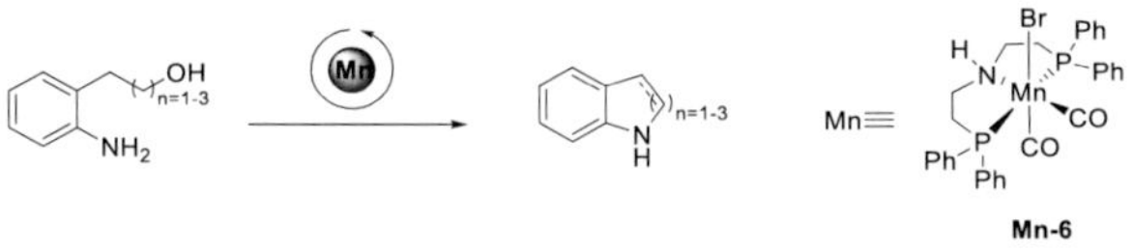

Scheme 44. Manganese-catalysed synthesis of heterocycles *via* intramolecular alkylation of amines with alcohols.

Many heterocyclic scaffolds are considered as privileged structures as bigger part of developed drugs consists in their core a heterocyclic fragment. Therefore, heterocycles play a central role in modern drug design.[162] As an example, vilazodone is a medication used to treat major depressive disorder, whilst oxamniquine serves as antihelminthic agent and tolvaptan is used to treat hyponatremia (low levels of sodium in blood) (Scheme 45).

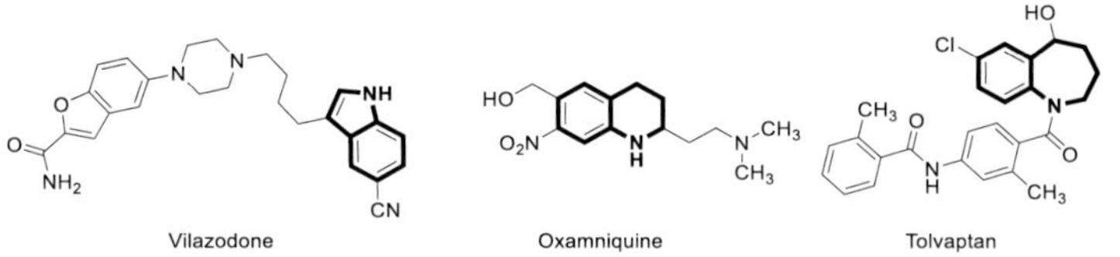

Scheme 45. Approved drugs containing heterocyclic fragment.

A strategy to build heterocyclic skeleton *via* oxidative cyclization of amino alcohols is not new. Significant amount of works was published featuring synthesis of *N*-containing heterocycles using this strategy. Vast majority of them focused on using rare noble metals including Ru,[163] Ir,[103,164,165] Pd,[165,166] Fe[167] and Ni.[168] Additionally, most of the above mentioned works were focused on intermolecular coupling of amines with alcohols and only mentioned indole synthesis from a commercial 2-(2-aminophenyl)ethan-1-ol. Lack of reports regarding application of base-metals for a synthesis of heterocycles using oxidative cyclisation strategy and a poor scope represented in known reports were my motivation to employ a Mn-catalyst which is active for (de)hydrogenation reactions.

Recently, significant attention was paid to the application of manganese, as the third most abundant transition metal in the Earth's crust, for the catalytic (de)hydrogenation of organic molecules.[83–88] A large number of reports were published featuring manganese-catalysed C-N bond formation,[55,61,72,169] including monoalkylation of amines with methanol.[72–74,170] Meanwhile, the strategy was also applied to a synthesis of a wide range of heterocycles.[58,63–65,110,111,171] To the best of my knowledge, a manganese-catalysed intramolecular alkylation of amines with alcohols was only mentioned in the work published by Beller's group[72] on the example of indole formation from a commercial 2-(2-aminophenyl)ethan-1-ol. Giving a high importance of heterocyclic compounds in medicinal chemistry (Scheme 45), a further development and scope expansion is highly desired.

2.6.2 Optimisation of reaction conditions

In the beginning, oxidative *N*-heterocyclization of 2-(2-aminophenyl)ethan-1-ol **24a** was investigated using **Mn-6** under various reaction conditions. Preliminary experiments were carried out in toluene in the presence of numerous bases with stirring for 16 hours at 120 °C (Table 13, entries 1-4). KO*t*Bu showed the best performance and 92% of the desired indole **25a** was formed. Next, different solvents were investigated. The reaction in dioxane provided an excellent yield of the desired indole (Table 13, entry 5). When THF and acetonitrile were applied respectively lower and slightly lower yields were observed (Table 13, entries 6 and 7). Due to a high price of dioxane I decided to use toluene for further optimisation. Running the reaction for 24 hours led to a slight increase in the yield, 96% of the compound **25a** was produced (Table 13, entry 8). Further concentration of the reaction mixture and decreasing a catalyst loading to 1 mol% allowed to achieve full conversion and an excellent yield of indole **25a** (Table 13, entries 9 and 10). Control experiments in the absence of the catalyst or a base led to a formation of insignificant amount of the product (Table 13, entries 11 and 12). It should be noted that trace amounts of indoline was formed together with a desired indole **25a**.

Table 13. Optimisation of the reaction conditions.[a]

entry	base (5 mol%)	solvent	yield of **25a**, (%) [b]
1	KO^{t}Bu	toluene, 0.25 M	92
2	CsOH*H$_2$O	toluene, 0.25 M	85
3	Cs$_2$CO$_3$	toluene, 0.25 M	78
4	K$_2$CO$_3$	toluene, 0.25 M	40
5	KO^{t}Bu	dioxane 0.25 M	>99
6	KO^{t}Bu	THF 0.25 M	44
7	KO^{t}Bu	acetonitril 0.25 M	90
8 [c]	KO^{t}Bu	toluene, 0.25 M	96
9 [c]	KO^{t}Bu	toluene, 0.5 M	>99
10 [c, d]	KO^{t}Bu	toluene, 0.5 M	>99
11 [e]	-	toluene, 0.5 M	7
12 [c,e]	KO^{t}Bu	toluene, 0.5 M	8

[a] Reaction conditions: **24a** (0.25 mmol), **Mn-6** (2 mol%), base (5 mol%), selected solvent at 120 °C for 16 h. [b] Determined by the GC analysis using dodecane as internal standard. [c] 24 h. [d] **Mn-6** (1 mol%), KO^{t}Bu (2.5 mol%). [e] No catalyst.

2.6.3 Scope of substrates

Having optimised reaction conditions various amino alcohols were submitted to an oxidative N-heterocyclization affording indole derivatives (Table 14). A range of different amino alcohols with a substitution on the aromatic ring were transformed into the corresponding indoles in high to excellent yields (Table 14).

Table 14. Synthesis of indoles catalysed by **Mn-6**.[a]

Entry	Substrate	Product	Yield
1	24a	25a	89%
2	24b	25b	98%
3	24c	25c	95%
4	24d	25d	98%
5	24e	25e	95%
6	24f	25f	85%

[a] Reaction conditions: **24** (0.25 mmol), **Mn-6** (1 mol%), KO^{t}Bu (2.5 mol%), toluene (0.5 M) at 120 °C for 24 h. Isolated yields are given.

Next, 3-(2-aminophenyl)propan-1-ol **26a** was tested under optimised reaction conditions, however, 1,2,3,4-tetrahydriquinoline **27a** was formed with a poor yield. After optimisation of reaction conditions, the reaction showed the best performance when 2 mol% of the catalyst and 1 equiv. of KO^{t}Bu were used and the reaction temperature had to be increased to 135 °C. Thus, 1,2,3,4-tetrahydroquinoline was observed as a single product in 94% isolated yield (Table 15, entry 1). Various amino alcohols with a substitution on the aromatic ring, including halogenated substrates were tolerated and showed excellent selectivity and yields of the tetrahydroquinolines (Table 15, entries 2-5). Additionally, 2,3,4,5-tetrahydro-1H-benzo[b]azepine could be synthesised with 84% isolated yield.

Table 15. Synthesis of *N*-heterocycles catalysed by **Mn-6**.[a]

Entry	Substrate	Product	Yield
1	26a	27a	94%
2	26b	27b	97%
3	26c	27c	98%
4	26d	27d	98%
5	26e	27e	95%
6	26f	27f	84%

[a] Reaction conditions: **26** (0.25 mmol), **Mn-6** (2 mol%), KO*t*Bu (1 equiv.) toluene (0.5 M) at 135 °C for 24 h. Isolated yields are given.

2.6.4 Reaction mechanism

Although the mechanism for the described reaction was not studied thoroughly, based on the manganese complexes behaviour and observed products, a possible mechanism is presented in a Scheme 46. First, in both cases (compound **24a** or **26a**) undergoes dehydrogenation of the alcohol leading to an aldehyde intermediate **A** or **B** respectively. Next, nucleophilic attack to carbonyl group occurs leading to a cyclisation with a release of a water molecule. In case of a starting compound **24a** the formed imine is tautomerized to indole **25a** and hydrogenated manganese catalyst would release a hydrogen molecule to close a catalytic cycle. In case of a starting compound **26a** formed imine **C** is rapidly hydrogenated with Mn-H$_2$ species to produce 1,2,3,4-tetrahydroquinoline **27a** and active **Mn-6**.

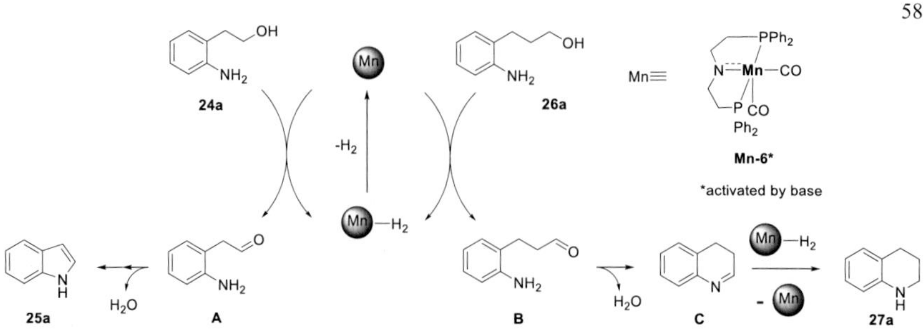

Scheme 46. Proposed reaction mechanism.

2.6.5 Summary

In conclusion, a new catalytic system for the synthesis of indoles, 1,2,3,4-tetrahydroquinolines, and 2,3,4,5-tetrahydro-1H-benzo[b]azepine was developed using well defined manganese pincer complex. The reaction proceeds under relatively mild reaction conditions leading to desired heterocycles in high to excellent yields. The displayed procedure allows an attractive and beneficial method for the synthesis of diverse N-heterocyclic compounds.

2.7 Manganese-catalysed alkylation of nitroarenes with alcohols

2.7.1 Introduction

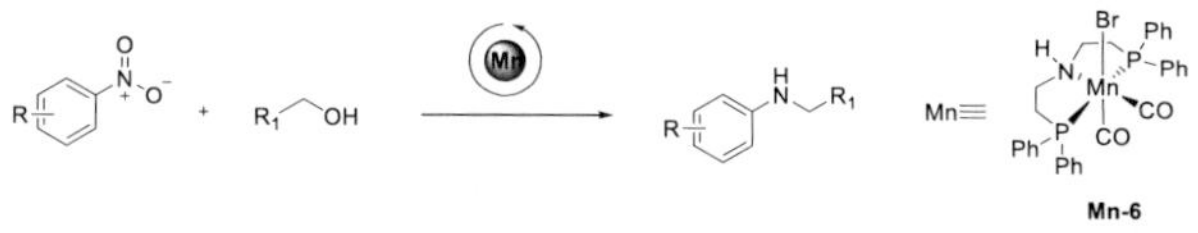

Scheme 47. Manganese-catalysed alkylation of nitroarenes with alcohols.

The formation of C-N bonds is one of the most essential synthetic reactions since nitrogen containing compounds are widely applied as agrochemicals, pharmaceutical drugs, dyes and polymers.[114] One of the biggest tasks of green chemistry is to create waste-free reactions using easily accessible and cheap reagents. An alkylation of amines with alcohols using transition metal catalysis is a promising approach to build C-N bonds which also satisfies demands of a green chemistry.[51,66,68,98,100,101,172] A way to improve this system is to use nitroarenes as starting materials, as they are cheaper and more available than amines. In this multistep reaction nitroarenes would be *in situ* reduced to an aniline which would undergo an alkylation. The strategy was previously studied predominantly using transition metal catalysts.[173] The substitution of noble-metal catalysts by earth-abundant alternatives is highly necessary.[37] In chapter 2.5 an efficient hydrogenation of nitroarenes with a use of **Mn-6** was described. Later in chapter 2.6, **Mn-6** proved to be reactive for alcohol dehydrogenation. Combining both procedures would lead to a direct alkylation of nitroarenes with alcohols (Scheme 47).

2.7.2 Optimisation of reaction conditions

Initially a manganese-catalysed alkylation of nitroarenes using methanol was studied. Thus *para*-methylnitrobenzene (**28b**) was selected as a benchmark substrate. Initially the reaction was tested using 3 mol% of the **Mn-6** and 20 mol% of base. Four different bases were investigated, leading to similar performance (Table 16, entries 1-4). Taking into account the small difference between the conversion and the yield of the desired product, Cs_2CO_3 was chosen for the following optimisation. Next, the amount of base was increased to 1 equiv. leading to a higher conversion and the additional formation of *p*-toluidine in 14% GC yield (Table 16, entry 5). Further increase of the amount of base led to the full conversion of the starting nitroarene with more than 99% of the desired alkylated aniline **30b** (Table 16, entry 6). A lower catalyst loading led to the full conversion of the starting material into 25% of *p*-toluidine and 61% of the desired alkylated product (Table 16, entry 7). The control experiment without the catalyst showed 55% conversion of the starting material without any formation of products. It indicates that due the use of an excess of base the starting material undergoes decomposition (Table 16, entry 8). The reaction without the base did not show any conversion (Table 16, entry 9). Thereafter, a substrate loading was increased to 0.5 mmol providing the same performance as with 0.25 mmol (Table 16, entry 6 and 10).

Table 16. Optimisation of reaction conditions.[a]

Me–C₆H₄–NO₂ (28b) + MeOH (0.25 M), Mn-6, base, 140 °C, 24 h → Me–C₆H₄–NH₂ (29b) + Me–C₆H₄–NH–CH₃ (30b)

entry	base	conv., %	yield of 29b (%)[b]	yield of 30b (%)[b]
1	KO^{t}Bu 20 mol%	17	-	14
2	Cs$_2$CO$_3$ 20 mol%	20	-	17
3	CsOH*H$_2$O 20 mol%	17	-	10
4	K$_2$CO$_3$ 20 mol%	20	-	14
5	Cs$_2$CO$_3$ 1 equiv.	60	14	17
6	Cs$_2$CO$_3$ 2 equiv.	full	-	>99
7[c]	Cs$_2$CO$_3$ 2 equiv.	full	25	61
8[d]	Cs$_2$CO$_3$ 2 equiv.	55	-	-
9	-	nr	-	-
10[e]	Cs$_2$CO$_3$ 2 equiv.	full	-	>99

[a] Reaction conditions: **28b** (0.25 mmol), **Mn-6** (3 mol%), base, MeOH (0.25 M) at 140 °C for 24 h. [b] Determined by the GC analysis using dodecane as internal standard. [c] **Mn-6** (2 mol%). [d] No catalyst. [e] **28b** (0.5 mmol).

2.7.3 Scope of substrates

With the optimised reaction conditions in hand, the substrates scope has been explored Unfortunately, only few substrates were tolerated, including unsubstituted phenyl and alkyl substituted nitroarenes (Table 17, **30a-c**). Compounds **30a-c** were isolated with high yields, as single products. Next, other alcohols were applied. Thus, only primary alcohols were tolerated leading to alkylated anilines in good yields (Table 17, **31a-c**). Attempts to use halogenated nitroarenes led to a nucleophilic aromatic substitution of the halogen by alkoxy group. Substrates with alkyl groups in *ortho* position were inactive in the described reaction. Reactions with secondary alcohols led to multiple products which were not further analysed (Scheme 48).

Table 17. Scope of substrates.[a]

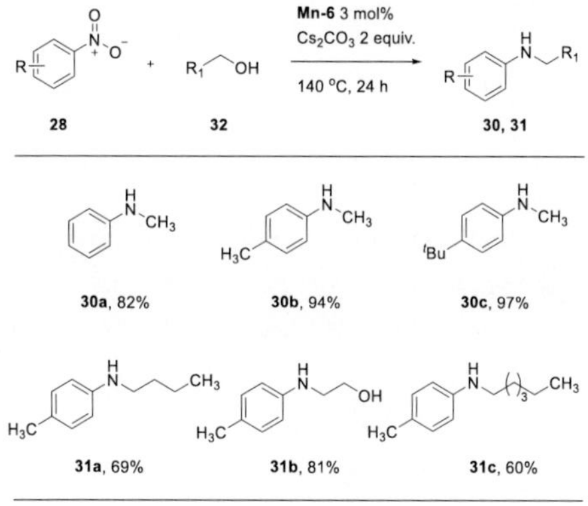

[Reaction scheme: nitroarene **28** + alcohol **32** with Mn-6 (3 mol%), Cs$_2$CO$_3$ (2 equiv.), 140 °C, 24 h → **30, 31**]

30a, 82%
30b, 94%
30c, 97%
31a, 69%
31b, 81%
31c, 60%

[a] Reaction conditions: nitrobenzene **28** (0.5 mmol), **Mn-6** (3 mol%), Cs$_2$CO$_3$ (2 equiv.), alcohol **32** (0.25 M) at 140 °C for 24 h.

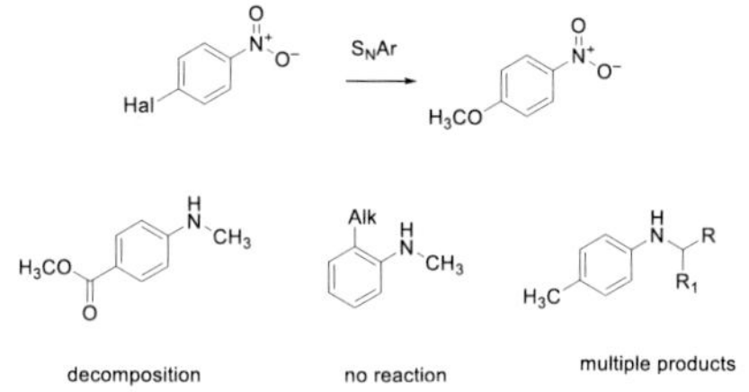

[Scheme showing limitations: Hal-substituted nitroarene → S$_N$Ar gives H$_3$CO-substituted product; methyl ester substrate → decomposition; ortho-alkyl substrate → no reaction; secondary amine substrate → multiple products]

Scheme 48. Scope limitations.

2.7.4 Summary

A manganese-catalysed alkylation of nitroarenes with alcohols was developed. Unfortunately, the substrates scope for this reaction was poor. Only alkyl substituted nitroarenes and primary alcohols were well tolerated. Due to the low performance of the developed reaction and the appearance of a similar research done by Morrill's group,[174] the investigations related to this project were suspended.

Chapter 3: Summary and Outlook

The presented projects in this thesis are based on the development of new catalytic systems using base metal catalysis. The work in this thesis addressed the synthesis of a wide range of important molecules and the establishment of sustainable procedures. Manganese-catalysed protocols, described in this work, satisfy demands of green chemistry and at the same time strongly compete with noble metal catalysed procedures.

In the first project, an air and moisture stable Mn-PNN complex **Mn-1** was applied for a selective hydrogenation of various CO_2 derived cyclic organic carbonates to diols and methanol. The procedure illustrates an indirect hydrogenation of CO_2 to methanol. Together with cyclic organic carbonates formamides and CO_2 derived polypropylene carbonate were successfully applied. **Mn-1** operates under low catalyst loading (up to 0.5 mol%) providing high yields of the products. Additionally, deuterated methanol (95% D) could be achieved when deuterium is used instead of hydrogen. The reaction mechanism was studied using DFT calculations revealing that the reaction proceeds *via* metal-ligand cooperative catalysis. An extension of this project can be the hydrogenation of other CO_2 derived compounds, such as ureas or carbamates.

In the next project I used the same manganese complex for asymmetric amination of racemic alcohols using hydrogen autotransfer strategy. The developed protocol tolerates a wide range of substrates producing desired amines in high yields with very good diastereomeric ratio. The produced chiral amines are used as intermediates for a synthesis of pharmaceuticals and fungicides. Notably, this environmentally benign, atom economic protocol uses readily available substrates producing water as sole by-product. Performed mechanistic studies suggest a metal-ligand cooperative catalysis. Further improvements in this project could be directed into the development of a new chiral manganese complex which can catalyse direct asymmetric amination of racemic alcohols.

In the third project manganese-catalysed semihydrogenation of alkynes using molecular hydrogen as reducing agent has been described. A newly developed **Mn-8** complex was successfully applied to selectively produce *(Z)*-alkenes under mild reaction conditions, although small amounts of over hydrogenated products were observed. The developed catalytic system shows good reactivity and tolerates a variety of functional groups and heterocycles leading to a practical synthesis of *(Z)*-olefins as well as allylic alcohols. Unfortunately, the developed reaction could not be applied for the hydrogenation of terminal alkynes. A gram scale experiment was performed highlighting a high potential of the developed reaction. An improvement for this project could be an optimisation of reaction conditions to produce alkanes from alkynes. Additionally, alkenes could be explored as starting substrates.

In the next project a single manganese catalyst was developed which can catalyse both, the hydrogenation and acceptorless dehydrogenation of *N*-containing heterocycles. The developed protocol is highly chemoselective and the products of both hydrogenation and dehydrogenation reactions can be isolated in good to excellent yields. The applied catalyst **Mn-6** is scalable, air and moisture stable and can be easily synthesised from readily available components. Mechanistic studies indicate a metal-ligand cooperative catalysis. The project can be potentially improved by the application of a broader substrate scope, including other *N*-containing heterocycles. Alternatively, a chiral manganese catalyst could be developed for an asymmetric hydrogenation of *N*-containing heterocycles.

In the fifth project, manganese-catalysed hydrogenation of nitroarenes was developed using molecular hydrogen as reducing agent. The desired aniline derivatives are synthesised in excellent yields at the same time tolerating a wide range of functional groups. The developed protocol was applied for a synthesis of an intermediate of a commercial drug vortioxetine, which is used to treat major depressive disorder. Additionally, a gram scale hydrogenation of *para*-iodonitrobenzene was performed leading to a corresponding aniline in a good yield. This result indicates a high feasibility of the developed procedure. The performed mechanistic studies suggested that the reaction proceeds *via* metal-ligand cooperative catalysis. As a follow up project, manganese-catalysed transfer hydrogenation of nitroarenes could be investigated.

In the next project, a manganese-catalysed protocol for the synthesis of 5, 6, and 7-membered N-containing heterocycles was developed. The desired heterocycles are produced as single products in high to excellent yields using relatively mild reaction conditions. The demonstrated procedure offers an attractive and beneficial method for the synthesis of diverse N-heterocyclic compounds. An intermolecular alkylation of aliphatic amines with alcohols could be suggested to improve the project, or other manganese catalysts could be tested in order to produce oxindoles or 3,4-dihydroquinolin-2(1H)-ones. Additionally, asymmetric cyclisation could be studied.

In the last project a manganese-catalysed alkylation of nitroarenes with alcohols, including methanol, was developed. Only alkyl substituted nitroarenes and primary alcohols were tolerated in the developed system. Unfortunately, due to significant scope limitations, investigations related to this project were suspended. The addition of an extra hydrogen donor could improve this protocol.

Chapter 4: Experimental part

4.1 Manganese-catalysed hydrogenation of cyclic organic carbonates

4.1.1 General information

All reactions were carried out under an argon atmosphere using oven-dried glassware. The dry and degassed dioxane and 2-methyl-THF were distilled from sodium benzophenone under nitrogen. THF, diethylether, toluene and hexane were obtained from MBRAUN Solvent Purification System. ^{1}H, ^{13}C and ^{31}P spectra were recorded in CDCl$_3$ or DMSO-$d6$ using Varian VNMR 600 MHz spectrometer. The signals were referenced to residual chloroform (7.26 ppm, ^{1}H, 77.00 ppm, ^{13}C). Chemical shifts are reported in ppm, multiplicities are indicated as s (singlet), d (doublet), t (triplet), br (broad), dd (doublet of doublets) and m (multiplet). IR spectra were recorded on a Perkin Elmer-100 spectrometer and are reported in terms of frequency of absorption (cm^{-1}). Mass spectra (EI-MS, 70 eV) were conducted on a Finnigan SSQ 7000 spectrometer. HRMS were recorded on a Thermo Scientific LTQ Orbitrap XL spectrometer. Analytical thin-layer chromatography (TLC) was performed using silica gel 60 pre-coated aluminium plates (Macherey-Nagel 0.20 mm thickness) with a fluorescent indicator UV254. Visualization was performed with standard phosphomolybdic acid stain (10g in 100 mL EtOH) or UV light. The single crystal XRD data were collected on SuperNova single crystal X-ray diffractometer equipped with a cooling device. GC conditions: CP-Sil-8-CB (30m, ID-1µm, df- 0.25µm) column, carrier gas: H$_2$, injection temp.: 250 °C, detector temp.: 300 °C, linear flow-75kPa, oven temp.: 50 °C, 5 min, 20 °C/min, 150 °C, 2 min, 40 °C/min, 250 °C. All chemicals if not otherwise noted were purchased and used without further purification. Carbonates **1c-1j, 1l** an **1m** were synthesised according reported procedure.[175]

4.1.2 Ligand and complex synthesis and characterization

2-(Diphenylphosphaneyl)-N-(pyridin-2-ylmethyl)ethan-1-amine (L1)[176]

A solution of 2-picolylaldehyde (0.514 g, 4.8 mmol) in THF (5 mL) was slowly added to a solution of the PN ligand (1.0 g, 4.36 mmol) in THF (20 mL). The reaction mixture was stirred for 1 h at room temperature. After the completion of the reaction the solvent was reduced in *vacuo* and the residue was dissolved in toluene (20 mL). A solution of diisobutylaluminium hydride in toluene (4.36 mL, 5.23 mmol, 1.2 M) was subsequently added to the reaction mixture dropwise. The reaction mixture was stirred for another 1 h at room temperature, and then quenched with water and extracted with toluene. The organic layer was dried with Na_2SO_4 and concentrated under reduced pressure. The residue was purified by column chromatography using ethyl acetate/Et_3N as eluent to afford **L1** as pale-yellow oil. (1.20 g, 86%).

^{1}H NMR (600 MHz, $CDCl_3$): δ 8.53 (d, J = 4.8 Hz, 1H), 7.60-7.50 (m, 1H), 7.43-7.40 (m, 4H), 7.3 (m, 6H), 7.24 (d, J = 7.8 Hz, 1H),7.14 (dd, J = 7.3, 5.0 Hz, 1H), 3.88 (s, 2H), 2.81 (dd, J = 15.6, 7.9 Hz, 2H), 2.34 – 2.31 (m, 2H), 2.04 (br. s, 1H). **^{13}C NMR** (151 MHz, $CDCl_3$): δ 159.5, 149.3, 138.3 (d, J = 12.2 Hz), 136.4, 132.7 (d, J = 18.8 Hz), 128.6, 128.5, 128.4 (d, J = 6.8 Hz), 122.2, 121.9, 55.0, 46.3 (d, J = 21.4 Hz), 29.0 (d, J = 11.9 Hz). **^{31}P NMR** (243 MHz, $CDCl_3$): δ -20.78.

Synthesis of [Mn(PhPNN)(CO)$_3$]Br (Mn-1)

A 50 mL Schlenk tube was charged with the PNN pincer ligand **L1** (300 mg, 1.07 mmol, 1 equiv.) and $Mn(CO)_5Br$ (293 mg, 1.07 mmol, 1.0 equiv.). The Schlenk tube was evacuated and backfilled with argon for several times. Afterwards, 15 mL of degassed THF was added and the reaction mixture was stirred at 80 °C for 20 h. The suspension was allowed to cool to room temperature and the yellow precipitate was filtered off and washed with diethyl ether and n-hexane. The remaining solid was dried under vacuum to afford the complex **Mn-1** as a yellow powder (0.49 g, 84% yield).

^{1}H NMR (600 MHz, DMSO-$d6$): δ 7.93 – 7.92 (m, 1H), 7.87 – 7.84 (m, 2H), 7.81-7.78 (m, 1H), 7.58-7.49 (m, 4H), 7.39 – 7.37 (m, 1H), 7.31 – 7.29 (m, 2H), 7.20 – 7.09 (m, 2H), 6.94 – 6.92 (m, 1H), 4.58 – 4.54 (m, 1H), 4.38 (m, 1H), 3.28 – 3.20 (m, 1H), 3.04 – 3.00 (m, 1H), 2.95-2.89 (m, 1H), 2.40 – 2.27 (m, 1H). **^{13}C NMR** (151 MHz, DMSO-$d6$): δ 221.2, 219.9, 215.3, 162.1, 153.0, 139.2, 132.4, 132.0, 131.1, 130.9, 130.8, 129.9, 129.8, 129.6, 125.0, 122.4, 59.8, 53.8 (d, J = 8.8 Hz), 22.7 (d, J = 22.7 Hz). **^{31}P NMR** (242 MHz, DMSO-$d6$): δ 65.86. **IR (ATR):** ν 3045, 2891, 2024, 1914, 1843, 1477, 1434, 1099, 892, 752, 693 cm^{-1}. **HRMS (ESI):** calc. for $C_{23}H_{21}MnN_2O_3P$ [M-Br]$^+$: 459.0665, found 459.0675.

Crystals suitable for X-ray diffraction were obtained in the following way. In an inner vial was placed the complex followed by toluene, DCM and triethylamine (the solution was filtered to get it clear) in the outer vial toluene was placed. Upon standing at room temperature for 48 h yellow crystals were formed. The supernatant solution was removed with a help of syringe and the crystals were immediately covered with PARATONE® and mounted on the diffractometer. The data were collected on SuperNova single crystal X-ray diffractometer equipped with a cooling device.

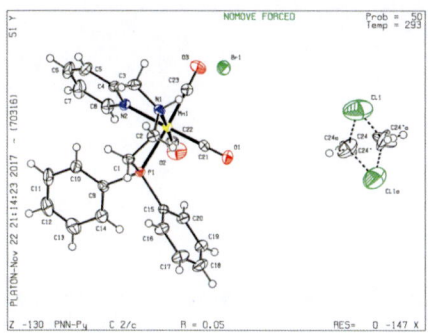

Figure 13. ORTEP plot of **Mn-1** with thermal ellipsoids at the 50% probability level.[5]

Table 18. Crystal data and structure refinement for **Mn-1**.

Identification code	**Mn-1**
Empirical formula	$C_{23.50}H_{22}BrClMnN_2O_3P$
Formula weight	581.70
Temperature	293(2) K
Wavelength	0.71073 Å
Crystal system	Monoclinic
Space group	C2/c
Unit cell dimensions	a = 30.4003(4) Å α = 90°
	b = 11.63168(12) Å β = 98.9262(12)°
	c = 13.86942(18) Å γ = 90°
Volume	4844.92(10) Å^3
Z	8
Density (calculated)	1.595 Mg/m^3
Absorption coefficient	2.399 mm^{-1}
F(000)	2344
Crystal size	0.550 x 0.420 x 0.330 mm^3
Theta range for data collection	3.427 to 30.277°.
Index ranges	-41<=h<=42, -16<=k<=16, -18<=l<=19
Reflections collected	265623
Independent reflections	6714 [R(int) = 0.1054]
Completeness to theta = 25.242°	99.7 %

---

[5] The crystal structure measurement was performed by Dr. Y. Lebedev

Absorption correction	Semi-empirical from equivalents
Max. and min. transmission	1.00000 and 0.67744
Refinement method	Full-matrix least-squares on F^2
Data / restraints / parameters	6714 / 30 / 307
Goodness-of-fit on F^2	1.086
Final R indices [I>2sigma(I)]	R1 = 0.0486, wR2 = 0.1378
R indices (all data)	R1 = 0.0551, wR2 = 0.1450
Extinction coefficient	n/a
Largest diff. peak and hole	0.968 and -1.666 e. Å^{-3}

Table 19. Bond lengths.

Bond	Length (Å)	Bond	Length (Å)
Mn(1)-C(22)	1.789(3)	C(5)-C(6)	1.379(6)
Mn(1)-C(21)	1.814(3)	C(6)-C(7)	1.376(6)
Mn(1)-C(23)	1.842(3)	C(7)-C(8)	1.386(4)
Mn(1)-N(2)	2.045(2)	C(9)-C(14)	1.392(4)
Mn(1)-N(1)	2.083(2)	C(9)-C(10)	1.397(4)
Mn(1)-P(1)	2.3335(7)	C(10)-C(11)	1.397(4)
P(1)-C(15)	1.824(2)	C(11)-C(12)	1.371(6)
P(1)-C(1)	1.828(3)	C(12)-C(13)	1.373(6)
P(1)-C(9)	1.830(3)	C(13)-C(14)	1.398(5)
O(1)-C(21)	1.141(4)	C(15)-C(16)	1.386(4)
O(2)-C(22)	1.153(4)	C(15)-C(20)	1.394(4)
O(3)-C(23)	1.137(4)	C(16)-C(17)	1.395(4)
N(1)-C(3)	1.485(4)	C(17)-C(18)	1.379(5)
N(1)-C(2)	1.490(3)	C(18)-C(19)	1.382(5)
N(2)-C(4)	1.344(4)	C(19)-C(20)	1.387(4)
N(2)-C(8)	1.353(4)	Cl(1)-C(24')	1.669(14)
C(1)-C(2)	1.519(4)	Cl(1)-C(24)	1.676(13)
C(3)-C(4)	1.494(4)	C(24)-Cl(1)#1	1.687(12)
C(4)-C(5)	1.402(4)	C(24')-Cl(1)#1	1.666(13)

Table 20. Bond angles.

Bond	Angle [°]	Bond	Angle [°]
C(22)-Mn(1)-C(21)	90.41(13)	N(1)-C(2)-C(1)	110.7(2)
C(22)-Mn(1)-C(23)	88.04(13)	N(1)-C(3)-C(4)	111.7(2)
C(21)-Mn(1)-C(23)	92.52(12)	N(2)-C(4)-C(5)	121.2(3)
C(22)-Mn(1)-N(2)	97.04(12)	N(2)-C(4)-C(3)	117.2(2)
C(21)-Mn(1)-N(2)	172.52(11)	C(5)-C(4)-C(3)	121.5(3)
C(23)-Mn(1)-N(2)	88.43(11)	C(6)-C(5)-C(4)	119.3(3)
C(22)-Mn(1)-N(1)	177.46(11)	C(7)-C(6)-C(5)	119.0(3)
C(21)-Mn(1)-N(1)	91.49(11)	C(6)-C(7)-C(8)	119.7(3)
C(23)-Mn(1)-N(1)	93.56(11)	N(2)-C(8)-C(7)	121.5(3)
N(2)-Mn(1)-N(1)	81.04(9)	C(14)-C(9)-C(10)	118.6(3)
C(22)-Mn(1)-P(1)	94.71(9)	C(14)-C(9)-P(1)	120.9(2)
C(21)-Mn(1)-P(1)	89.84(8)	C(10)-C(9)-P(1)	120.5(2)
C(23)-Mn(1)-P(1)	176.37(9)	C(11)-C(10)-C(9)	120.3(3)
N(2)-Mn(1)-P(1)	88.88(6)	C(12)-C(11)-C(10)	120.2(3)
N(1)-Mn(1)-P(1)	83.62(6)	C(11)-C(12)-C(13)	120.3(3)
C(15)-P(1)-C(1)	107.88(12)	C(12)-C(13)-C(14)	120.2(3)
C(15)-P(1)-C(9)	103.91(11)	C(9)-C(14)-C(13)	120.3(3)
C(1)-P(1)-C(9)	105.02(12)	C(16)-C(15)-C(20)	119.3(2)
C(15)-P(1)-Mn(1)	115.80(8)	C(16)-C(15)-P(1)	122.3(2)
C(1)-P(1)-Mn(1)	101.15(8)	C(20)-C(15)-P(1)	118.35(19)
C(9)-P(1)-Mn(1)	121.92(8)	C(15)-C(16)-C(17)	119.6(3)
C(3)-N(1)-C(2)	112.3(2)	C(18)-C(17)-C(16)	120.7(3)
C(3)-N(1)-Mn(1)	110.81(17)	C(17)-C(18)-C(19)	119.9(3)
C(2)-N(1)-Mn(1)	114.65(16)	C(18)-C(19)-C(20)	119.8(3)
C(4)-N(2)-C(8)	119.2(2)	C(19)-C(20)-C(15)	120.7(3)
C(4)-N(2)-Mn(1)	115.61(19)	O(1)-C(21)-Mn(1)	179.6(3)
C(8)-N(2)-Mn(1)	125.1(2)	O(2)-C(22)-Mn(1)	176.7(3)
C(2)-C(1)-P(1)	108.15(18)	O(3)-C(23)-Mn(1)	173.0(3)

2-(diphenylphosphanyl)-N-methyl-N-(pyridin-2-ylmethyl)ethan-1-amine (L2)

i) HCHO, DCE, 15 min
ii) NaBH(OAc)$_3$, 24h, rt

L1 → L2

Aqueous formaldehyde (1.70 g, 20.90 mmol, 33% in water) was added to a solution of bis[2-(2-pyridyl)methyl]amine (2.08 g, 10.45 mmol) in 1,2-dichloroethane (50 mL). After 15 min, NaBH(OAc)$_3$ (4.42 g, 20.90 mmol) was added portion-wise to the reaction mixture and stirring was continued for 24 h at room temperature. The reaction was quenched by the addition of an aqueous solution of 2 M NaOH (100 mL). The organic layer was separated and the aqueous layer was extracted with DCM. The organic layer was dried with Na$_2$SO$_4$ and concentrated under reduced pressure. The residue was purified by column chromatography using gradient hexane/ethyl acetate/Et$_3$N mixture as eluent to afford the pure methylated ligand L2 (370 mg, 47% yield).

^{1}**H NMR** (600 MHz, CDCl$_3$): δ 8.51 (d, J = 4.5 Hz, 1H), 7.65-7.54 (m, 1H), 7.45-7.38 (m, 4H), 7.37-7.35 (m, 1H), 7.34-7.27 (m, 6H), 7.16-7.07 (m, 1H), 3.66 (s, 2H), 2.67-2.51 (m, 2H), 2.33 – 2.27 (m, 2H), 2.27 (s, 3H). ^{13}**C NMR** (151 MHz, CDCl$_3$): δ 159.1, 149.1, 138.4 (d, J = 12.8 Hz), 136.4, 132.7 (d, J = 18.7 Hz), 128.6, 128.4 (d, J = 6.7 Hz), 123.1, 121.9, 63.3, 54.1(d, J = 22.4 Hz), 42.1, 25.9 (d, J = 11.9 Hz). ^{31}**P NMR** (243 MHz, CDCl$_3$): δ -19.98.

Synthesis of [Mn(PhPNN)(CO)3]Br (Mn-1 (N-Me))

L2 + Mn(CO)$_5$Br → Mn-1(N-Me), 88% yield
ethanol, 60 °C

A 50 mL Schlenk tube was charged with the Me-PNN pincer ligand L2 (100 mg, 0.3 mmol, 1.2 equiv.) and Mn(CO)5Br (68 mg, 0.25 mmol, 1.0 equiv.). The Schlenk tube was evacuated and backfilled with argon for several time. Afterwards, 10 mL of degassed ethanol was added and the reaction mixture was stirred at 60 °C for 16 h. The suspension was then allowed to cool to room temperature and concentrated under reduced pressure. The remaining solid was washed with diethyl ether and n-hexane and dried under vacuum to afford the complex Mn-1(N-Me) as a yellow powder (122 mg, 88% yield).

^{1}**H NMR** (600 MHz, DMSO-$d6$) δ 8.05 – 7.72 (m, 4H), 7.72 – 7.43 (m, 4H), 7.43 – 7.04 (m, 5H), 6.00 – 5.98 (m, 1H), 4.56 (d, J = 15.2 Hz, 1H), 4.39 (d, J = 15.9 Hz, 1H), 3.57 – 3.31 (m, 1H), 3.29 – 2.91 (m,

5H), 1.12 – 0.96 (m, 1H). ^{13}C NMR (151 MHz, DMSO-$d6$): δ 221.2, 218.5, 212.9, 159.3, 153.1, 139.0, 132.1 (d, J = 10.7 Hz), 131.7, 130.8, 130.5, 130.0 (d, J = 10.0 Hz), 129.4 (d, J = 10.5 Hz), 129.2 (d, J = 10.1 Hz), 128.5, 128.2, 69.4, 60.5, 54.0, 25.8 (d, J = 24.2 Hz). ^{31}P NMR (242 MHz, DMSO-$d6$): δ 64.00.

4.1.3 General procedures and reaction analysis

General procedure for the hydrogenation of cyclic organic carbonates

A 20 mL glass vial was charged with the cyclic organic carbonate **1** (1 mmol), **Mn-1** (5.4 mg, 0.01 mmol), KO^{t}Bu (2.8 mg, 0.025 mmol) and 4 mL of degassed dioxane. The sealed vial was transferred into a stainless steel autoclave and a hole was made with a needle to allow an access of the gases. The autoclave was carefully flushed three times with nitrogen and then hydrogen gas. After adjusting the final hydrogen pressure to 50 bar, the autoclave was heated to 140 °C for 16 h with stirring. The residual H$_2$ was released carefully and the mixture was analysed by GC with m-xylene as an internal standard. Furthermore, the crude reaction mixture was purified by column chromatography on silica gel to obtain the pure diol. Each reaction was typically repeated several times and the average results are reported.

Hydrogenative depolymerization of poly(propylene carbonate) (PPC)

A 20 mL glass vial was charged with poly(propylene carbonate) (100 mg, 1 mmol), **Mn-1** (10.8 mg, 0.02 mmol), KO^{t}Bu (5.6 mg, 0.05 mmol) and 4 mL of degassed dioxane. The sealed vial was transferred into a stainless steel autoclave and a hole was made with a needle to allow an access of the gases. The autoclave was carefully flushed three times with nitrogen and then hydrogen gas. After adjusting the final hydrogen pressure to 50 bar, the autoclave was heated to 140 °C for 16 h with stirring. The residual H$_2$ was released carefully in and the mixture was analysed by GC with m-xylene as an internal standard.

Hydrogenation of morpholine-4-carbaldehyde

A 20 mL glass vial was charged with morpholine-4-carbaldehyde **4** (115 mg, 1 mmol), **Mn-1** (2.7 mg, 0.005 mmol), KO^{t}Bu (1.4 mg, 0.0125 mmol) and 4 mL of degassed dioxane. The sealed vial was transferred into a stainless steel autoclave and a hole was made with a needle to allow an access of the gases. The autoclave was carefully flushed three times with nitrogen and then hydrogen gas. After adjusting the final hydrogen pressure to 50 bar, the autoclave was heated to 140 °C for 16 h with stirring. The residual H$_2$ was released carefully and the mixture was analysed by GC with m-xylene as an internal standard.

Deuterium-labeling experiments

A glass vial was charged with **1a** or **2a** (4 mmol), **Mn-1** (43.2 mg, 0.08 mmol), KO^{t}Bu (22.4 mg, 0.2 mmol) and 16 mL of degassed dioxane. The sealed vial was transferred into a stainless steel autoclave and a hole was made with a needle to allow an access of the gases. The autoclave was carefully flushed three times with nitrogen. After adjusting the final deuterium pressure to 50 bar, the autoclave was heated to 140 °C for 16 h with stirring. After cooling down to room temperature, the residual D_2 was released carefully. Afterwards, 20 mL of DCM, pyridine (1.2 mL) and benzoyl chloride (1.6 mL) were added to the crude reaction mixture at 0 °C, and the resulting mixtures were stirred at room temperature for 18 h. The solvent was removed under vacuum and the residue was purified by column chromatography on silica gel using a mixture of petroleum ether and ethyl acetate (50/1 ~ 30/1) as the eluent, to afford methyl benzoate and ethylene dibenzoate, respectively. The deuterium incorporation was determined using fourier-transform ion cyclotron resonance mass spectroscopy (FTICR).

FTICR spectroscopy of the deuteration of ethylene carbonate[6]

a) D-distribution in ethylene dibenzoate

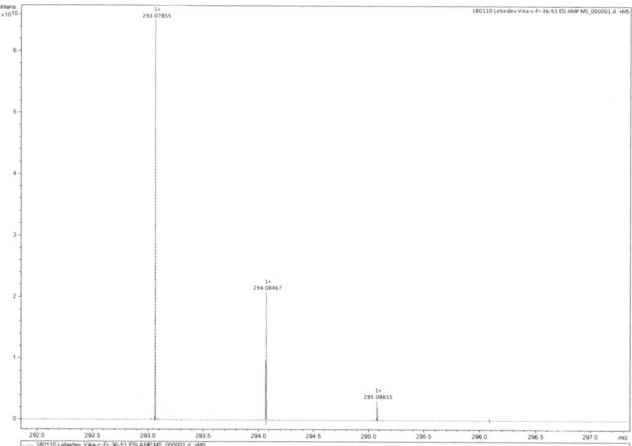

---

[6] The experiments were performed by Dr. Y. Lebedev

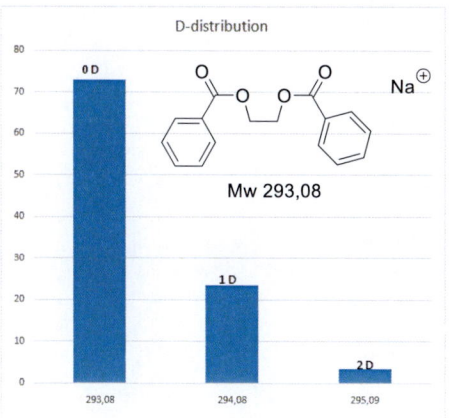

b) D-distribution in methyl benzoate

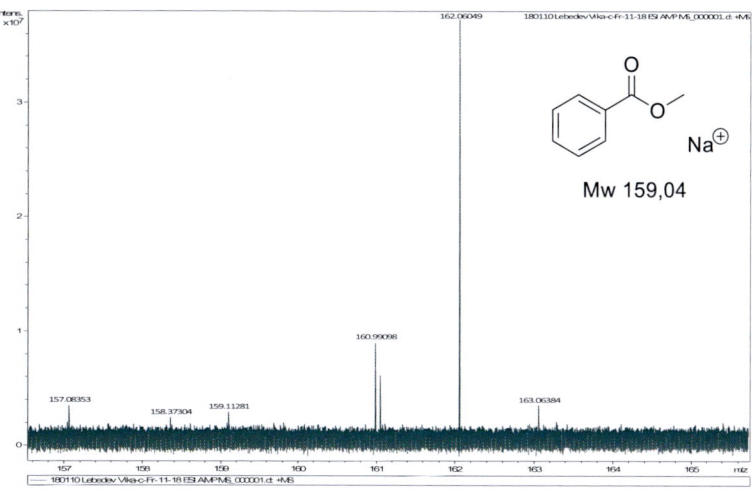

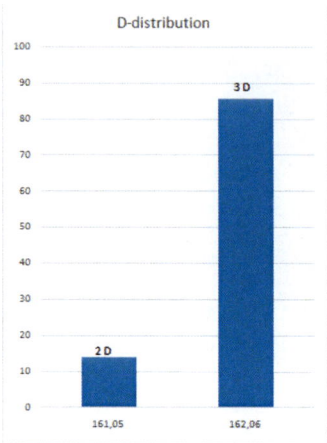

FTICR spectroscopy of the deuteration of ethylene glycol

a) D-distribution in ethylene dibenzoate

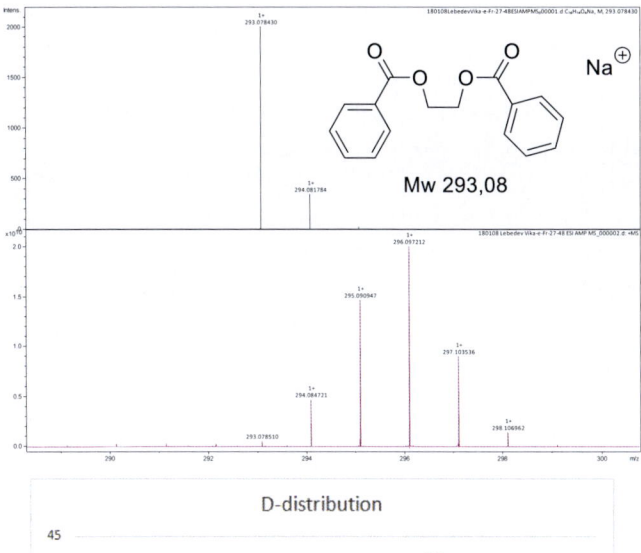

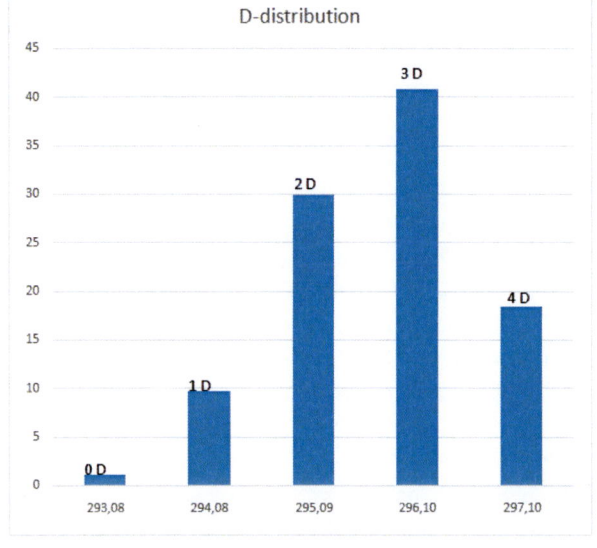

Hydrogenation of paraformaldehyde

A 20 mL glass vial was charged with paraformaldehyde (30 mg, 1 mmol), **Mn-1** (10.8 mg, 0.02 mmol), KO^{t}Bu (5.6 mg, 0.05 mmol) and 4 mL of degassed dioxane. The sealed vial was transferred into a stainless steel autoclave and a hole was made with a needle to allow an access of the gases. The autoclave was carefully flushed three times with nitrogen and then hydrogen gas. After adjusting the final hydrogen pressure to 50 bar, the autoclave was heated to 140 °C for 16 h with stirring. The residual H_2 was released carefully in and the mixture was analysed by GC with m-xylene as an internal standard.

Competitive studies

A 20 mL glass vial was charged with **Mn-1** (5.4 mg, 0.01 mmol), KO^{t}Bu (2.8 mg, 0.025 mmol), the cyclic carbonate (**1f**, 72 mg, 0.5 mmol) and a 2:1 mixture of 2-hydroxy-3,3-dimethylbutyl formate and 1-hydroxy-3,3-dimethylbutan-2-yl formate (73 mg, 0.5 mmol). The degassed dioxane (4 mL) was subsequently added. The sealed vial was transferred into a stainless steel autoclave and a hole was made with a needle to allow an access of the gases. The autoclave was carefully flushed three times with nitrogen and then hydrogen gas. After adjusting the final hydrogen pressure of 50 bar, the autoclave was heated to 140 °C for 8 h with stirring. After cooling down the autoclave to room temperature, the residual H_2 was released carefully and the mixture was analysed by GC with m-xylene as an internal standard.

Hexane-1,2-diol (2d)[19]

96%, colorless oil. **^{1}H NMR** (600 MHz, CDCl$_3$) δ 3.69 – 3.63 (m, 1H), 3.61-3.59 (m, 1H), 3.52-3.47 (m, 2H), 3.4-3.37 (m, 1H), 1.44 – 1.36 (m, 3H), 1.36 – 1.23 (m, 3H), 0.88 (t, J = 7.1 Hz, 3H).**^{13}C NMR** (151 MHz, CDCl$_3$) δ 72.3, 66.7, 32.7, 27.7, 22.7, 14.0.

Octane-1,2-diol (2e)[177]

99%, colorless oil. **^{1}H NMR** (600 MHz, CDCl$_3$): δ 3.75 – 3.56 (m, 2H), 3.44-3.41 (m, 1H), 2.74-2.40 (br, 2H), 1.42 (s, 3H), 1.31-1.27 (m, 7H), 0.89-0.86 (m, 3H). **^{13}C NMR** (151 MHz, CDCl$_3$): δ 72.4, 66.8, 33.1, 31.7, 29.3, 25.5, 22.6, 14.1.

3,3-dimethylbutane-1,2-diol (2f)[178]

99%, colorless oil. **^{1}H NMR** (400 MHz, CDCl$_3$) δ 3.73-3.69 (m, 1H), 3.48-3.43 (m, 1H), 3.37-3.34 (m, 1H), 2.75-2.73 (m, 2H), 0.89 (s, 9H).**^{13}C NMR** (101 MHz, CDCl$_3$) δ 79.7, 63.1, 33.5, 25.8.

1-Phenylethane-1,2-diol (2g)[19]

99%, white solid. **^{1}H NMR** (600 MHz, CDCl$_3$): δ 7.37 – 7.30 (m, 5H), 4.81-4.79 (m, 1H), 3.74-3.71 (m, 1H), 3.64-3.62 (m, 1H), 2.92 (s, 2H). **^{13}C NMR** (151 MHz, CDCl$_3$): δ 140.4, 128.5, 128.0, 126.1, 74.7, 68.1.

3-(Benzyloxy)propane-1,2-diol (2h)[19]

99%, colorless oil. **^{1}H NMR** (600 MHz, CDCl$_3$): δ 7.37-7.29 (m, 5H), 4.56 (s, 2H), 3.94 – 3.56 (m, 5H), 2.80-2.43 (br, 2H). **^{13}C NMR** (151 MHz, CDCl$_3$): δ 137.6, 128.5, 127.9, 127.8, 73.6, 72.0, 70.7, 64.2.

3-Methoxypropane-1,2-diol (2i)[19]

99%, colorless oil. **^{1}H NMR** (400 MHz, CDCl$_3$): δ 3.88-3.76 (m, 1H), 3.65-3.51 (m, 2H), 3.44 – 3.37 (m, 3H), 3.34 (s, 3H). **^{13}C NMR** (101 MHz, CDCl$_3$): δ 74.1, 70.6, 63.9, 59.2.

Butane-2,3-diol (2j)[19]

99%, *(R,S)/meso* =1.5/1 colorless oil. **^{1}H NMR** (600 MHz, CDCl$_3$): δ 3.83 – 3.70 (m, 1.2 H, *chiral*), 3.55 – 3.43 (m, 0.8 H, *meso*), 3.20-2.30 (br, 2H), 1.20-1.03 (m, 6H). **^{13}C NMR** (151 MHz, CDCl$_3$): δ 72.5 (meso), 70.8 (*R,S*), 19.2 (meso), 16.8 (*R,S*).

3-methylbutane-1,3-diol (2l)[179]

99%, colorless oil. **^{1}H NMR** (600 MHz, CDCl$_3$): δ 4.06 – 3.98 (m, 1H), 3.86-3.72 (m, 2H), 3.57-3.34 (m, 2H), 1.71-1.59 (m, 2H), 1.24-1.14 (m, 3H). **^{13}C NMR** (151 MHz, CDCl$_3$) δ 67.9, 61.3, 39.9, 23.6.

Butane-1,3-diol (2m)[180]

99%, colorless oil. **^{1}H NMR** (600 MHz, CDCl$_3$): δ 3.85 (t, *J* = 5.7 Hz, 2H), 3.27 (br, 2H), 1.70 (t, *J* =5.8 Hz, 2H), 1.26 (s, 6H). **^{13}C NMR** (151 MHz, CDCl$_3$): δ 71.9, 59.9, 43.1, 29.5.

4.2 Diastereoselective amination of racemic alcohols

4.2.1 General information

All reactions were carried out under an argon atmosphere using oven-dried glassware. The dry and degassed dioxane and 2-methyl-THF were distilled from sodium benzophenone under nitrogen. The dry and degassed *t*-amyl alcohol was distilled from calcium hydride under nitrogen. Toluene were obtained from MBRAUN Solvent Purification System. All other chemicals were used as purchased without further purification. ^{1}H, ^{13}C and ^{19}F spectra were recorded in CDCl$_3$ using Varian VNMR 600 MHz and 400MHz spectrometer. The signals were referenced to residual chloroform (7.26 ppm, ^{1}H, 77.00 ppm, ^{13}C). Chemical shifts are reported in ppm, multiplicities are indicated by s (singlet), d (doublet), t (triplet), dd (doublet of doublets), qd (quartet of doublets), br (broad signal) and m (multiplet). IR spectra were recorded on a Perkin Elmer-100 spectrometer and are reported in terms of frequency of absorption (cm^{-1}). Mass spectra (EI-MS, 70 eV) were conducted on a Finnigan SSQ 7000 spectrometer. HRMS were recorded on a Thermo Scientific LTQ Orbitrap XL spectrometer. Analytical thin-layer chromatography (TLC) was performed using silica gel 60 pre-coated aluminium plates (Macherey-Nagel 0.20 mm thickness) with a fluorescent indicator UV254. Visualization was performed with standard phosphomolybdic acid stain (10g in 100 mL EtOH) or UV light.

4.2.2 General procedures and reaction analysis

General procedure for the asymmetric amination of secondary alcohols.

$$\text{R}^1\text{-CH(OH)-R}^2 \; (\textit{rac}\text{-}7) + \text{H}_2\text{N-S(O)-}^t\text{Bu} \; ((R_s)\text{-}8) \xrightarrow[\text{140 °C, 16 h, TAA 0.5 M}]{\textbf{Mn-1}, \text{Cs}_2\text{CO}_3} \text{HN(S(O)}^t\text{Bu})\text{-CH(R}^1\text{)-R}^2 \; ((R,R_s)\text{-}9) + \text{H}_2\text{O}$$

An oven dried 25 mL Schlenk tube equipped a stir bar was charged with secondary alcohol **7** (0.75 mmol), (R)-2-methyl-2-propanesulfinamide (61 mg, 0.5 mmol), **Mn-1** (5-10 mol%) and Cs_2CO_3 (10-20 mol%). Then the tube was evacuated and backfilled with argon for three times and degassed *t*-amyl alcohol (1 mL) was added. The reaction mixture was stirred at 140 °C in aluminium block for 16-48 h. Upon cooling down to room temperature, the residue was directly purified by flash column chromatography using silica gel and EtOAc/hexanes mixtures as eluent to give the pure *N*-alkylated sulfinamide (*R*, *R*s)-**9**.

Diastereomeric ratio was determined by ^{1}H NMR analysis of crude reaction mixture. For the known compounds, the data was compared with the literature reports.[107,108,181] For the other compounds, authentic samples of the (*S*, *R*s) diastereoisomers were prepared by the reduction of the corresponding sulfinamide imine with L-selectride.[181]

(R)-2-methyl-*N*-((R)-1-phenylethyl)propane-2-sulfinamide ((R, Rs)-9a) [107]

85%, colorless oil. **^{1}H NMR:** (600 MHz, CDCl$_3$) δ 7.37-7.25 (m, 4H), 7.29-7.26 (m, 1H), 4.54 (qd, *J*=6.5, 2.6 Hz, 1H), 3.43 (br, 1H), 1.50 (d, *J* = 6.6 Hz, 3H), 1.22 (s, 9H). **^{13}C NMR:** (151 MHz, CDCl$_3$) δ 144.0, 128.7, 127.8, 126.6, 55.4, 53.8, 53.8, 22.8, 22.6. **MS (EI):** m/z = 225.0 [M]$^+$. **IR (ATR):** ν = 3208, 2972, 2325, 2086, 1453, 1365, 1310, 1201, 1181, 1055, 919, 851, 762, 699cm^{-1}.

(R)-*N*-((R)-1-(4-chlorophenyl)ethyl)-2-methylpropane-2-sulfinamide ((R, Rs)-9b) [182]

80%, colorless solid. **^{1}H NMR:** (600 MHz, CDCl$_3$) δ 7.28 – 7.24 (m, 4H), 4.49 – 4.46 (m, 1H), 3.40 (d, *J* = 3.4 Hz, 1H), 1.45 (d, *J* = 6.7 Hz, 3H), 1.18 (s, 9H). **^{13}C NMR:** (151 MHz, CDCl$_3$) δ 142.5, 133.4, 128.8, 128.0, 55.5, 53.4, 53.4, 22.8, 22.5. **MS (EI):** m/z = 259.9 [M]$^+$. **IR (ATR):** ν =3278, 2980, 1488, 1417, 1365, 1289, 1203, 1177, 1121, 1087, 1045, 938, 820, 783, 717, 658 cm^{-1}.

(R)-N-((R)-1-(4-fluorophenyl)ethyl)-2-methylpropane-2-sulfinamide ((R, R$_s$)-9c) [107]

65%, colorless solid. **¹H NMR:** (600 MHz, CDCl$_3$) δ 7.30 – 7.28 (m, 2H), 6.99 (t, J = 8.7 Hz, 2H), 4.52 – 4.48 (m, 1H), 3.38 (br, 1H), 1.46 (d, J = 6.7 Hz, 3H), 1.20 (s, 9H). **¹³C NMR:** (151 MHz, CDCl$_3$) δ 163.0, 161.3, 139.8 (d, J = 3.0 Hz), 128.2 (d, J = 8.1 Hz), 115.6, 115.5, 55.5, 53.4 (d, J = 1.8 Hz), 22.9, 22.5. **¹⁹F NMR:** (282 MHz, CDCl$_3$) δ -114.5. **MS (EI):** m/z = 243.0 [M]⁺. **IR (ATR):** ν = 3213, 2978, 2956, 2925, 2868, 1603, 1509, 1456, 1423, 1363, 1281, 1219, 1160, 1119, 1088, 1043, 941, 863, 830, 733 cm⁻¹.

(R)-N-((R)-1-(3-chlorophenyl)ethyl)-2-methylpropane-2-sulfinamide ((R, R$_s$)-9d) [107]

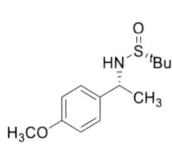

74%, white solid. **¹H NMR:** (400 MHz, CDCl$_3$) δ 7.31 – 7.27 (m, 1H), 7.26 – 7.16 (m, 3H), 4.47 (qd, J = 6.5, 3.2 Hz, 1H), 3.41 (d, J = 2.1 Hz, 1H), 1.45 (d, J = 6.6 Hz, 3H), 1.19 (s, 9H). **¹³C NMR:** (101 MHz, CDCl$_3$) δ 146.0, 134.5, 130.0, 127.9, 126.6, 124.9, 55.5, 53.6, 22.8, 22.5. **MS (EI):** m/z = 259.9 [M]⁺. **IR (ATR):** ν =3160, 2960, 1579, 1433, 1366, 1204, 1125, 1041, 947, 879, 832, 781, 693 cm⁻¹.

(R)-N-((R)-1-(4-methoxyphenyl)ethyl)-2-methylpropane-2-sulfinamide ((R, Rs)-9e) [107]

87%, white solid. **¹H NMR:** (600 MHz, CDCl$_3$) δ 7.24 (d, J = 8.6 Hz, 2H), 6.83 (d, J = 8.6 Hz, 2H), 4.47 (qd, J = 6.4, 2.3 Hz, 1H), 3.75 (s, 3H), 3.36 (s, 1H), 1.45 (d, J = 6.6 Hz, 3H), 1.19 (s, 9H). **¹³C NMR:** (151 MHz, CDCl$_3$) δ 159.1, 136.1, 127.7, 114.0, 55.3, 55.2, 53.3, 22.7, 22.6. **MS (EI):** m/z = 255.0 [M]⁺. **IR (ATR):** ν =3216, 2963, 2324, 2083, 1611, 1512, 1458, 1366, 1304,1281, 1244, 1177, 1044, 913, 830, 731, 683, 662 cm⁻¹.

(R)-N-((R)-1-(4-(tert-butyl)phenyl)ethyl)-2-methylpropane-2-sulfinamide ((R, R$_s$)-9f)

85%, colorless oil. **¹H NMR:** (600 MHz, CDCl$_3$) δ 7.35 (d, J = 8.4 Hz, 2H), 7.27 (d, J = 8.4 Hz, 2H), 4.52 (qd, J = 6.5, 2.6 Hz, 1H), 3.42 (d, J = 1.0 Hz, 1H), 1.48 (d, J = 6.7 Hz, 3H), 1.30 (s, 9H), 1.21 (s, 9H). **¹³C NMR:** (151 MHz, CDCl$_3$) δ 150.6, 141.0, 126.3, 125.6, 55.4, 53.5, 34.5, 31.3, 22.6. **MS (EI):** m/z = 281.0 [M]⁺. IR (ATR): ν = 3287, 2960, 2867,1511, 1469, 1421, 1365, 1271, 1208, 1119, 1041, 946, 829 cm⁻¹. **HRMS (ESI):** calc. for C$_{16}$H$_{28}$ONS [M + H]⁺: 282.1886, found 282.1889.

(R)-N-((R)-1-(3-methoxyphenyl)ethyl)-2-methylpropane-2-sulfinamide ((R, R$_s$)-9g) [182]

83%, yellow oil. **¹H NMR:** (400 MHz, CDCl$_3$) δ 7.22-7.26 (m, 1H), 6.94 – 6.86 (m, 2H), 6.80-6.82 (m, 1H), 4.50 (qd, J = 6.4, 2.7 Hz, 1H), 3.79 (s, 3H), 3.41 (s, 1H), 1.48 (d, J = 6.5 Hz, 3H), 1.22 (s, 9H). **¹³C NMR:** (101 MHz, CDCl$_3$) δ 159.8, 145.7, 129.8, 118.8, 112.9, 112.4, 55.4, 55.2, 53.8, 22.7, 22.6. **MS (EI):** m/z = 256.1 [M+H]⁺. **IR (ATR):** ν = 3214, 2968, 2326, 2162, 1711, 1598, 1458, 1365, 1313, 1256, 1165, 1047, 940, 874, 784, 699 cm⁻¹.

(R)-N-((R)-1-(benzo[d][1,3]dioxol-5-yl)ethyl)-2-methylpropane-2-sulfinamide ((R, R_s)-9h)[182]

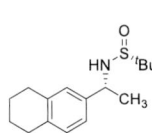

95%, light yellow solid. ¹H NMR: (600 MHz, CDCl₃) δ 6.80-6.77 (m, 1H), 6.76-6.73 (m, 1H), 6.71-6.68 (m, 1H), 5.88 (s, 2H), 4.41 (qd, J = 6.5, 2.7 Hz, 1H), 3.36 (d, J = 1.6 Hz, 1H), 1.41 (d, J = 6.7 Hz, 3H), 1.17 (s, 9H). ¹³C NMR: (151 MHz, CDCl₃) δ 147.8, 147.0, 138.0, 119.9, 108.2, 106.8, 101.0, 55.4, 53.7, 22.8, 22.6. **MS (EI):** m/z = 269.9 [M]⁺. **IR (ATR):** ν = 3304, 2970, 2879, 1610, 1485, 1441, 1376, 1305, 1236, 1097, 1040, 920, 815, 728, 671 cm⁻¹.

(R)-2-methyl-N-((R)-1-(5,6,7,8-tetrahydronaphthalen-2-yl)ethyl)propane-2-sulfinamide ((R, R_s)-9i)

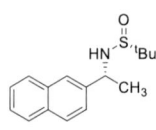

70%, white solid. ¹H NMR: (400 MHz, CDCl₃) δ 7.08 – 6.98 (m, 3H), 4.46 (qd, J = 6.5, 2.5 Hz, 1H), 3.35 (s, 1H), 2.74-2.73 (m, 4H), 1.81 – 1.74 (m, 4H), 1.47 (d, J = 6.5 Hz, 3H), 1.21 (s, 9H). ¹³C NMR: (101 MHz, CDCl₃) δ 141.1, 137.4, 136.8, 129.5, 127.3, 123.6, 55.3, 53.6, 29.4, 29.1, 23.1, 22.6. **MS (EI):** m/z = 280.2 [M+H]⁺. **IR (ATR):** ν = 3227, 2920, 1435, 1365, 1162, 1135, 1036, 948, 812, 706 cm⁻¹. **HRMS (ESI):** calc. for C₁₆H₂₆ONS [M + H]⁺: 280.1730, found 280.1730.

(R)-2-methyl-N-((R)-1-(naphthalen-2-yl)ethyl)propane-2-sulfinamide ((R, R_s)-9j) [107]

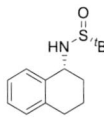

76%, light yellow solid. ¹H NMR: (600 MHz, CDCl₃) δ 7.88 – 7.75 (m, 4H), 7.51 – 7.44 (m, 3H), 4.72 (qd, J = 6.5, 2.8 Hz, 1H), 3.55 (d, J = 1.4 Hz, 1H), 1.60 (d, J = 6.6 Hz, 3H), 1.25 (s, 9H). ¹³C NMR: (151 MHz, CDCl₃) δ 141.3, 133.3, 133.0, 128.6, 128.0, 127.7, 126.3, 126.1, 125.3, 124.7, 55.5, 54.0, 22.6. **MS (EI):** m/z = 275.2 [M]⁺. **IR (ATR):** ν =3254, 2979, 2953, 2900, 2867, 2113, 2078, 1983, 1925, 1727, 1603, 1509, 1466, 1363, 1307, 1177, 1059, 1015, 913, 866, 821, 750, 674 cm⁻¹.

(R)-2-methyl-N-((R)-1,2,3,4-tetrahydronaphthalen-1-yl)propane-2-sulfinamide ((R, R_s)-9k) [107]

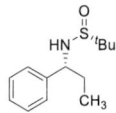

84%, yellow solid. ¹H NMR: (600 MHz, CDCl₃) δ 7.46 – 7.42 (m, 1H), 7.20 – 7.15 (m, 2H), 7.10 – 7.07 (m, 1H), 4.56-4.54 (m, 1H), 3.26 (d, J = 3.1 Hz, 1H), 2.83 – 2.77 (m, 1H), 2.75 – 2.68 (m, 1H), 2.02 – 1.95 (m, 1H), 1.95 – 1.84 (m, 2H), 1.78 – 1.71 (m, 1H), 1.20 (s, 9H). ¹³C NMR: (151 MHz, CDCl₃) δ 137.7, 136.9, 129.6, 129.2, 127.5, 126.5, 55.4, 52.8, 30.6, 29.1, 22.6, 18.2. **MS (EI):** m/z = 251.0 [M]⁺. **IR (ATR):** ν = 3188, 2935, 2859, 2325, 1726, 1456, 1361,1289, 1190, 1028, 898, 742 cm⁻¹.

(R)-2-methyl-N-((R)1-phenylpropyl)propane-2-sulfinamide ((R, R_s)-9l) [107]

80%, white solid. ¹H NMR: (400 MHz, CDCl₃) δ 7.40 – 7.16 (m, 5H), 4.28-4.24 (m, 1H), 3.37 (d, J = 2.4 Hz, 1H), 2.16 – 1.92 (m, 1H), 1.85 – 1.62 (m, 1H), 1.21 (s, 9H), 0.77 (t, J = 7.4 Hz, 9H). ¹³C NMR: (101 MHz, CDCl₃) δ 142.2, 128.6, 127.7, 127.2, 60.3, 55.6, 29.3, 22.6, 10.0. **MS (EI):** m/z = 239.1 [M]⁺. **IR (ATR):** ν = 3193, 2929, 2867, 1602, 1455, 1362, 1185, 1047, 893, 752, 696 cm⁻¹.

(R)-2-methyl-N-((R)-1-phenylpentyl)propane-2-sulfinamide ((R, R_s)-9m) [181]

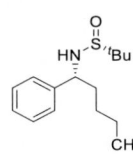

86%, colorless oil. **^{1}H NMR:** (600 MHz, CDCl$_3$) δ 7.37 – 7.26 (m, 5H), 4.34-4.31 (m, 1H), 3.38 (d, J = 3.1 Hz, 1H), 2.07 – 1.96 (m, 1H), 1.76 – 1.67 (m, 1H), 1.33-1.17 (m, 3H), 1.22 (s, 9H), 1.12 – 1.01 (m, 1H), 0.83 (t, J = 7.2 Hz, 3H). **^{13}C NMR:** (151 MHz, CDCl$_3$) δ 142.6, 128.6, 127.8, 127.2, 59.1, 55.6, 36.2, 27.8, 22.6, 22.5, 14.0. **MS (EI):** m/z = 267.1 [M]$^+$. **IR (ATR):** ν = 3215, 2939, 2867, 2326, 2100, 1887, 1743, 1601, 1459, 1368, 1185, 1052, 916, 703 cm^{-1}.

(R)-N-((R)-1-cyclohexylethyl)-2-methylpropane-2-sulfinamide ((R, R_s)-9n)

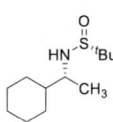

55%, colorless oil. **^{1}H NMR:** (600 MHz, CDCl$_3$) δ 3.23 – 3.16 (m, 1H), 3.10 (d, J = 4.2 Hz, 1H), 1.77 – 1.59 (m, 5H), 1.40 – 1.32 (m, 1H), 1.26 – 1.17 (m, 2H), 1.16 (s, 9H), 1.12-1.09 (m, 1H), 1.07 (d, J = 6.6 Hz, 3H), 1.04 – 0.93 (m, 2H). **^{13}C NMR:** (151 MHz, CDCl$_3$) δ 55.5, 55.3, 44.2, 29.3, 27.9, 26.4, 26.3, 26.2, 22.6, 17.8. **MS (EI):** m/z = 231.0 [M]+. **IR (ATR):** ν = 3238, 3176, 2920, 2851, 1730, 1447, 1365, 1152, 1043, 957, 906, 846 cm–1. **HRMS (ESI):** calc. for C$_{12}$H$_{26}$ONS [M + H]+: 232.1730, found 232.1729.

(R)-N-((R)-1-cyclopentylethyl)-2-methylpropane-2-sulfinamide ((R, R_s)-9o)

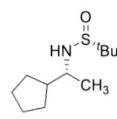

74%, colorless oil. **^{1}H NMR:** (400 MHz, CDCl$_3$) δ 3.26 – 3.17 (m, 1H), 3.13 (d, J = 2.6 Hz, 1H), 1.91 – 1.46 (m, 7H), 1.37 – 1.20 (m, 2H), 1.18 (s, 9H), 1.15 (d, J = 6.3 Hz, 3H). **^{13}C NMR:** (101 MHz, CDCl$_3$) δ 57.8, 55.2, 55.1, 46.8, 29.5, 29.3, 25.6, 25.4, 22.5, 20.1. **MS (EI):** m/z = 217.1 [M]$^+$. **IR (ATR):** ν = 3456, 3219, 2949, 2869, 2326, 2156, 1736, 1647, 1457, 1369, 1321, 1180, 1128, 1050, 909, 847, 796 cm^{-1}. **HRMS (ESI):** calc. for C$_{11}$H$_{23}$ONS [M + Na]$^+$: 240.1393, found 240.1389.

(R)-N-((R)-1-cyclopropylethyl)-2-methylpropane-2-sulfinamide ((R, R_s)-9p)

89%, colorless oil. **^{1}H NMR:** (400 MHz, CDCl$_3$) δ 3.23 (s, 1H), 2.54 – 2.44 (m, 1H), 1.18 (d, J = 6.4 Hz, 3H), 1.13 (s, 9H), 0.83 – 0.72 (m, 1H), 0.52 – 0.41 (m, 2H), 0.28 – 0.13 (m, 2H). **^{13}C NMR:** (101 MHz, CDCl$_3$) δ 56.2, 55.0, 22.4, 20.9, 18.8, 4.1, 3.5. **MS (EI):** m/z = 189.0 [M]$^+$. **IR (ATR):** ν = 3461, 3215, 2964, 2338, 2094, 1722, 1538, 1460, 1370, 1311, 1187, 1051, 949, 819, 682 cm^{-1}. **HRMS (ESI):** calc. for C$_9$H$_{19}$ONNaS [M + Na]$^+$: 212.1080, found 212.1082.

(R)-2-methyl-N-((R)-1-phenylpropan-2-yl)propane-2-sulfinamide ((R, R_s)-9q) [183]

78%, colorless oil. **^{1}H NMR:** (600 MHz, CDCl$_3$) δ 7.36 – 7.17 (m, 5H), 3.67 – 3.57 (m, 1H), 3.00 (d, J = 6.5 Hz, 1H), 2.84 (dd, J = 13.5, 6.8 Hz, 1H), 2.70 (dd, J = 13.5, 6.7 Hz, 1H), 1.29 (d, J = 6.5 Hz, 3H), 1.11 (s, 9H). **^{13}C NMR:** (151 MHz, CDCl$_3$) δ 138.3, 129.5, 128.3, 126.4, 55.7, 53.8, 44.5, 22.6, 22.5. **MS (EI):** m/z = 240.1 [M+H]$^+$. **IR (ATR):** ν =3211, 2965, 1605, 1456, 1367, 1320, 1131, 1047, 960, 878, 744, 699 cm^{-1}.

(R)-2-methyl-N-((R)-1-(thiophen-2-yl)ethyl)propane-2-sulfinamide ((R, R$_s$)-9r)

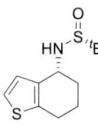
86%, yellow oil. **^{1}H NMR:** (600 MHz, CDCl$_3$) δ 7.23-7.22 (m, 1H), 7.03 (d, J = 3.4 Hz, 1H) 6.95-6.94 (m, 1H), 4.86-4.82 (m, 1H), 3.51 (d, J = 1.7 Hz, 1H), 1.59 (d, J = 6.6 Hz, 3H), 1.22 (s, 9H). **^{13}C NMR:** (151 MHz, CDCl$_3$) δ 147.9, 126.7, 124.7, 124.5, 55.6, 50.1, 23.8, 22.6. **MS (EI):** m/z = 231.9 [M]$^+$. **IR (ATR):** v = 3202, 2967, 2160, 1660, 1456, 1371, 1302, 1231, 1181, 1051, 908, 843, 698 cm^{-1}. **HRMS (ESI):** calc. for C$_{10}$H$_{18}$ONS$_2$ [M + H]$^+$: 232.0824, found 232.0828.

(R)-2-methyl-N-((R)-4,5,6,7-tetrahydrobenzo[b]thiophen-4-yl)propane-2-sulfinamide ((R, R$_s$)-9s)

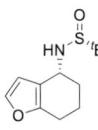
78%, white solid. **^{1}H NMR:** (400 MHz, CDCl$_3$) δ 7.12 – 7.06 (m, 2H), 4.53-4.50 (m, 1H), 3.30 (d, J = 4.7 Hz, 1H), 2.85 – 2.68 (m, 2H), 2.04 – 1.80 (m, 4H), 1.22 (s, 9H). **^{13}C NMR:** (101 MHz, CDCl$_3$) δ 138.9, 135.9, 127.3, 122.7, 55.5, 51.0, 31.2, 24.8, 22.6, 19.9. **MS (EI):** m/z = 257.0 [M]$^+$. **IR (ATR):** v = 3216, 2935, 2859, 1567, 1459, 1431, 1359, 1298, 1251, 1182, 1141, 1083, 1031, 929, 898, 870, 717, 674 cm^{-1}. **HRMS (ESI):** calc. for C$_{12}$H$_{19}$ONS$_2$ [M + Na]$^+$: 280.0800, found 280.0797.

(R)-2-methyl-N-((R)-4,5,6,7-tetrahydrobenzofuran-4-yl)propane-2-sulfinamide ((R, R$_s$)-9t)

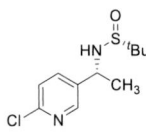
83%, white solid. **^{1}H NMR:** (400 MHz, CDCl$_3$) δ 7.24 (d, J = 1.3 Hz, 1H), 6.47 (d, J = 1.8 Hz, 1H), 4.40 (dd, J = 8.8, 4.7 Hz, 1H), 3.22 (d, J = 5.1 Hz, 1H), 2.63-2.49 (m, 2H), 1.99 – 1.74 (m, 4H), 1.20 (s, 9H). **^{13}C NMR:** (101 MHz, CDCl$_3$) δ 152.4, 141.0, 118.6, 109.6, 55.5, 49.5, 31.5, 22.8, 22.6, 19.3. **MS (EI):** m/z = 241.1 [M]$^+$. **IR (ATR):** v =3227, 2940, 2864, 1739, 1623, 1432, 1360, 1224, 1184, 1141, 1035, 955, 903, 842, 723 cm^{-1}. **HRMS (ESI):** calc. for C$_{12}$H$_{26}$ONS [M + Na]$^+$: 264.1029, found 264.1028.

(R)-N-((R)-1-(6-chloropyridin-3-yl)ethyl)-2-methylpropane-2-sulfinamide ((R, R$_s$)-9u)

65%, white solid. **^{1}H NMR:** (400 MHz, CDCl$_3$) δ 8.36 (d, J = 2.4 Hz, 1H), 7.65 (dd, J = 8.2, 2.5 Hz, 1H), 7.30 (d, J = 8.2 Hz, 1H), 4.55 (qd, J = 6.7, 3.9 Hz, 1H), 3.41 (d, J = 2.8 Hz, 1H), 1.53 (d, J = 6.6 Hz, 3H), 1.21 (s, 9H).**^{13}C NMR:** (101 MHz, CDCl$_3$) δ 148.2, 138.2, 137.3, 124.3, 55.8, 51.5, 22.6, 22.5. **MS (EI):** m/z = 261.0 [M+H]$^+$. **IR (ATR):** v =3217, 2962, 1733, 1575, 1458, 1373, 1283, 1204, 1099, 1029, 940, 841, 797 cm^{-1}. **HRMS (ESI):** calc. for C$_{11}$H$_{18}$ON$_2$ClS [M + H]$^+$: 261.0823, found 261.0820.

4.3 Hydrogenation of alkynes and alkenes catalysed by manganese pincer complex

4.3.1 General information

All reactions were carried out under an argon atmosphere using an oven-dried glassware. The dry and degassed methanol and *t*-amyl alcohol were distilled from calcium hydride under nitrogen. THF, DMF, toluene, hexane and DCM were obtained from MBRAUN Solvent Purification System. ^{1}H, ^{13}C and ^{31}P spectra were recorded in CDCl$_3$ or C$_6$D$_6$ using Varian VNMR 600 MHz or Inova 400 MHz spectrometer.

The signals were referenced to residual chloroform (7.26 ppm, ^{1}H, 77.00 ppm, ^{13}C). Chemical shifts are reported in ppm, multiplicities are indicated by s (singlet), d (doublet), t (triplet), q (quartet), dd (doublet of doublets), td (triplet of doublets), ddd (doublet of doublet of doublets), br (broad signal) and m (multiplet). IR spectra were recorded on a Perkin Elmer-100 spectrometer and are reported in terms of frequency of absorption (cm^{-1}). Mass spectra (EI-MS, 70 eV) were conducted on a Finnigan SSQ 7000 spectrometer. HRMS were recorded on a Thermo Scientific LTQ Orbitrap XL spectrometer. Analytical thin-layer chromatography (TLC) was performed using silica gel 60 pre-coated aluminium plates (Macherey-Nagel 0.20 mm thickness) with a fluorescent indicator UV254. Visualization was performed with standard phosphomolybdic acid stain (10g in 100 mL EtOH) or UV light. Analysis by gas chromatography (GC) was done using a CP-Sil-8-CB column (30 m, d = 0.25 mm) with FID detector and H$_2$ as carrier gas. All chemicals if not otherwise noted were purchased and used without further purification. **Mn-6 (N-Me)** was prepared following a procedure described in our previous report.[184] Starting alkynes were synthesized using Sonogashira cross-coupling method.[185]

4.3.2 Ligand and complex synthesis and characterization

L3 was prepared following slightly changed procedure.[186] **3-(2-chloroethyl)oxazolidin-2-one.** *N, N*-bis (chloroethyl) amine hydrochloride (20 g, 112 mmol, 1.0 equiv.), triethylamine (Et$_3$N) (31.3 mL, 224 mmol, 2 equiv.) and methanol (MeOH) (500 mL) were placed in a 500mL round bottom flask equipped with a magnetic stir bar. Carbon dioxide (CO$_2$) gas was generated from dry ice and bubbled into the obtained solution at room temperature for 1 hour. After concentration of the reaction solution under reduced pressure, toluene (800 mL) was added and the resulting white suspension was filtered and the residue was washed with toluene. The filtrates were combined and concentrated under reduced pressure to give 13.7 g of the title compound as a pale-yellow liquid. Isolated yield: 82%. Obtained 3-(2-chloroethyl)oxazolidin-2-one was used in the next step without further purification. ^{1}**H NMR** (400 MHz, CDCl$_3$) δ 4.38 – 4.31 (m, 2H), 3.76 – 3.63 (m, 4H), 3.62 – 3.56 (m, 2H).^{13}**C NMR** (101 MHz, CDCl$_3$) δ 62.0, 46.2, 45.7, 42.1.

3-(2-(Diphenylphosphaneyl)ethyl)oxazolidin-2-one. Under argon atmosphere a 200 mL Schlenk tube was charged with diphenylphosphine (6.7 g, 36 mmol, 1.1 equiv.) and 40 mL of dry THF. The solution was cooled to 5 °C in an ice-water bath. A hexane solution of *n*-butyllithium (22.5 mL, 36 mmol, 1.1 equiv.) was added dropwise by syringe. The temperature of the reaction mixture was kept below 10 °C. After the addition the ice-water bath was removed and the mixture was stirred for 20 minutes at room temperature to form lithium diphenylphosphide as a reddish liquid. (2- chloroethyl) -2- oxazolidinone (2) (4.9 g, 32.7 mmol, 1.0 equiv.) was placed in another Schlenk tube under argon atmosphere, next 60 mL of dry THF was added and the mixture was cooled to -30 °C using dry ice-acetone bath. Meanwhile, the solution of lithium diphenylphosphide was cooled to -20 °C and equipped with transferring cannula under argon atmosphere. It was added dropwise to the solution of (2- chloroethyl) -2- oxazolidinone over 2 hours. The resulting reaction mixture was further allowed to warm to room temperature. Next toluene 200 mL and water 50 ml were added, and it was concentrated under reduced pressure. Organic layer was separated and washed 3 times with 50 mL of water, dried over $MgSO_4$ and concentrated in vacuo. Obtained product was recrystallized from 2-methyl-2-buthanol. 5.5 g of 3-(2-(diphenylphosphaneyl)ethyl)oxazolidin-2-one was obtained as a white powder in 56% yield. ^{1}H NMR (600 MHz, CDCl$_3$) δ 7.50 – 7.27 (m, 10H), 4.19 – 4.10 (m, 2H), 3.52 – 3.46 (m, 2H), 3.42 (td, J = 9.3, 7.7 Hz, 2H), 2.36 – 2.32 (m, 2H). ^{13}C NMR (151 MHz, CDCl$_3$) δ 158.1, 137.5, 137.5, 132.7, 132.6, 128.9, 128.6, 128.6, 61.7, 44.7, 41.7, 41.6, 26.6, 26.5. ^{31}P NMR (243 MHz, CDCl$_3$) δ -21.3.

2-(diphenylphosphaneyl)-N-(2-(methylthio)ethyl)ethan-1-amine. (L3) A round bottom flask was charged with 3-(2-(diphenylphosphaneyl)ethyl)oxazolidin-2-one (3) (1 g, 3.34 mmol) and NaSMe (0.23 g, 3.34 mmol). 10 mL of 2-methyl-2-buthanol (*t*-amyl alcohol) was added, and the mixture was refluxed for 3 hours. The resulted mixture was concentrated under reduced pressure and then purified by column chromatography. Ethyl acetate with 10% NEt$_3$ were used as eluent. 0.84 g of 2-(diphenylphosphaneyl)-N-(2-(methylthio)ethyl)ethan-1-amine was obtained as pale-yellow viscous liquid giving 83% yield. ^{1}H NMR (600 MHz, CDCl$_3$) δ 7.45 – 7.40 (m, 4H), 7.36 – 7.29 (m, 6H), 2.80-2.75 (m, 4H), 2.61 (t, J = 6.5 Hz, 2H), 2.32 – 2.26 (m, 2H), 2.07 (s, 3H), 1.76 (br, 1H). ^{13}C NMR (151 MHz, CDCl$_3$) δ 138.3, 138.2, 132.8, 132.6, 128.6, 128.5, 128.4, 47.6, 46.3, 46.2, 34.2, 29.0, 28.9, 15.3. ^{31}P NMR (243 MHz, CDCl$_3$) δ -20.7.

Synthesis of manganese complex Mn-8

A 50 mL Schlenk tube was charged with Mn(CO)$_5$Br (453 mg, 1.65 mmol, 1.0 equiv.) and evacuated. Afterwards, 15 mL of degassed toluene was added under the flow of argon. Further, PNS pincer ligand **L3** (500 mg, 1.65 mmol, 1 equiv.) was added and the Schlenk tube was evacuated and backfilled with argon for several time to release formed CO. After that, the reaction mixture was stirred at 115 °C for 20 h. The suspension was allowed to cool to room temperature and the yellow precipitate was filtered off and washed with diethyl ether and n-hexane. The remaining solid was dried under vacuum to afford the complex **Mn-**

8 as a yellow powder (0.73 g, 85% yield). **^{1}H NMR** (600 MHz, CDCl$_3$) δ 7.85 (s, 2H), 7.42-7.29 (m, 8H), 3.73 (br, 2H), 3.40 (br, 1H), 3.17 (br, 1H), 2.99 – 2.54 (m, 4H), 2.47 (s, 1H), 2.41 (s, 3H). **^{13}C NMR** (151 MHz, CDCl$_3$) δ 138.1, 137.8, 134.5, 134.2, 133.6 (d, J = 9.3 Hz), 130.0, 129.7 (d, J = 9.6 Hz), 129.2, 128.5 (d, J = 9.1 Hz), 128.2 (d, J = 9.3 Hz), 52.1, 50.9, 37.2, 30.2 (d, J = 17.9 Hz), 23.1. **^{31}P NMR** (243 MHz, CDCl$_3$) δ 81.3. **HRMS (ESI):** calc. for C$_{20}$H$_{22}$O$_3$NMnPS [M-Br]$^+$: 442.0434, found 442.0433. **IR (ATR)** ν (cm^{-1}): 3157, 1918, 1832, 1428, 1299, 1179, 1098, 1072, 1007, 963, 843, 788, 742, 689.

4.3.3 General procedures and reaction analysis

In an argon filled glovebox a 15 mL glass vial was charged with the corresponding alkyne **11** (0.5 mmol), **Mn-8** (1-5 mol%), KO^{t}Bu (2.5-12.5 mol%) and 1 mL of degassed toluene. The vial was sealed with a cap with a septum and was transferred into a stainless-steel autoclave and a hole was made with a needle to allow an access of the gases. The autoclave was carefully flushed three times with nitrogen and then hydrogen gas. After adjusting the final hydrogen pressure to 20-30 bar, the autoclave was heated to 50-60 °C for 8-24h with stirring. After cooling down to room temperature the residual H$_2$ was vented carefully and was analysed by GC. Next the reaction mixture was filtered through syringe filter washed with ethyl acetate and evaporated. One equivalent of CH$_2$Br$_2$ was added and the mixture was further analysed by NMR. Furthermore, the crude reaction mixture was purified by column chromatography on silica gel to obtain the pure Z-alkenes.

Mn-1 catalysed hydrogenation of diphenylacetylene in a scale-up reaction
In an argon filled glovebox a 15 mL glass vial was charged with the corresponding alkyne **11a** (1 g, 5.61 mmol), **Mn-8** (14.6 mg, 0.5 mol%), KO^{t}Bu (7.87 mg, 1.25 mol%) and 5 mL of degassed toluene. The vial was sealed with a cap with a septum and transferred into a stainless-steel autoclave and a hole was made with a needle to allow an access of the gases. The autoclave was carefully flushed three times with nitrogen and then hydrogen gas. After adjusting the final hydrogen pressure to 20 bar, the autoclave was heated to 60 °C for 16h with stirring. After cooling down to room temperature the residual H$_2$ was vented carefully and was analysed by GC. Next the reaction mixture was filtered through syringe filter washed with ethyl acetate and evaporated and was further analysed by NMR. The reaction resulted in formation of Z-stylbene with 97% purity.

Mercury poison test

In an argon filled glovebox a 15 mL glass vial was charged with the diphenylacetylene **11a** (178 mg, 1 mmol), **Mn-8** (5.22 mg, 1 mol%), KO^{t}Bu (2.8mg, 2.5 mol%) and 2 mL of degassed toluene. The vial was sealed with a cap with a septum. A drop of mercury was added into a vial *via* syringe and the vial was transferred into a stainless-steel autoclave and a hole was made with a needle to allow an access of the gases. The autoclave was carefully flushed three times with nitrogen and then hydrogen gas. After adjusting the final hydrogen pressure to 20 bar, the autoclave was heated to 60 °C for 16 h with stirring. After cooling down to room temperature the residual H$_2$ was vented carefully and the mixture was analysed by GC. Next the reaction mixture was filtered through syringe filter, washed with ethyl acetate and evaporated. The residues of mercury were quenched with FeCl$_3$ solution and disposed. One equivalent of CH$_2$Br$_2$ was added and the mixture was further analysed by NMR giving full conversion of starting material and 95:1:4 ratio of **12a:13a:14a**

Stoichiometric reaction between Mn-1 and diphenylacetylene

In an argon filled glovebox a J. Young NMR tube was charged with **Mn-8** complex (15.7 mg, 0.03 mmol), KO^{t}Bu (8.43 mg, 0.075 mmol) and 0.7 mL of dry and degassed C$_6$D$_6$. The tube was sealed and placed in a sand bath with a temperature of 60 °C for 16 h. NMR of reaction mixture was measured and diphenylacetylene (5.4 mg, 0.03 mmol) was added in glovebox. The sample was afterwards placed in a sand bath with a temperature of 60 °C.

Spectral Data

(Z)-1,2-diphenylethene (12a)[187]

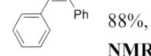

88%, colorless oil. **¹H NMR** (600 MHz, CDCl$_3$) δ 7.28 – 7.17 (m, 10H), 6.61 (s, 2H). **¹³C NMR** (151 MHz, CDCl$_3$) δ 137.2, 130.2, 128.9, 128.2, 127.1.

(Z)-1-methyl-4-styrylbenzene (12b)[187]

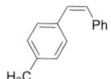

81%, colorless oil. **¹H NMR** (600 MHz, CDCl$_3$) δ 7.33 – 7.20 (m, 5H), 7.19 (d, *J* = 8.1 Hz, 2H), 7.07 (d, *J* = 7.9 Hz, 2H), 6.73 – 6.47 (m, 2H), 2.35 (s, 3H). **¹³C NMR** (151 MHz, CDCl$_3$) δ 137.5, 136.9, 134.3, 130.2, 129.6, 128.9, 128.9, 128.8, 128.2, 127.0, 21.3.

(Z)-1-methyl-2-styrylbenzene (12c)[188]

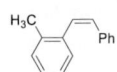

79%, colorless oil. **¹H NMR** (600 MHz, CDCl$_3$) δ 7.27 – 7.13 (m, 8H), 7.09 (t, *J* = 7.4 Hz, 1H), 6.67 (q, *J* = 12.2 Hz, 2H), 2.31 (s, 3H). **¹³C NMR** (151 MHz, CDCl$_3$) δ 137.1, 137.0, 136.1, 130.5, 130.1, 129.5, 128.9, 128.9, 128.1, 127.2, 127.0, 125.7, 19.9.

(Z)-1-ethyl-4-styrylbenzene (12d)[189]

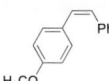

94%, colorless oil. **¹H NMR** (600 MHz, CDCl₃) δ 7.31 – 7.20 (m, 5H), 7.19 (d, *J* = 8.1 Hz, 2H), 7.06 (d, *J* = 8.1 Hz, 2H), 6.60 – 6.54 (m, 2H), 2.62 (q, *J* = 7.6 Hz, 2H), 1.23 (t, *J* = 7.6 Hz, 3H). **¹³C NMR** (151 MHz, CDCl₃) δ 143.2, 137.5, 134.5, 130.2, 129.5, 128.8, 128.2, 127.7, 126.9, 28.6, 15.4. **MS (EI)**: m/z = 208.7 [M]⁺. **IR (ATR)**: ν = 3466, 3015, 2964, 2924, 2867, 1956, 1904, 1601, 1509, 1448, 1383, 1182, 1117, 1067, 1025, 964, 918, 872, 831, 769, 696, 567, 521, 465 cm⁻¹. **HRMS (ESI)**: calc. for C₁₆H₁₆ [M]⁺: 208.1246, found 208.1246

(Z)-1-methoxy-4-styrylbenzene (12e)[187]

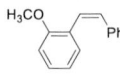

83%, white solid. **¹H NMR** (600 MHz, CDCl₃) δ 7.30 – 7.17 (m, 7H), 6.79 – 6.74 (m, 2H), 6.57 – 6.49 (m, 2H), 3.79 (s, 3H). **¹³C NMR** (151 MHz, CDCl₃) δ 158.6, 137.6, 130.1, 129.7, 129.6, 128.8, 128.7, 128.2, 126.9, 113.6, 55.2.

(Z)-1-methoxy-2-styrylbenzene (12f)[190]

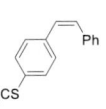

90%, colorless oil. **¹H NMR** (600 MHz, CDCl₃) δ 7.27 – 7.14 (m, 7H), 6.90 (d, J = 8.3 Hz, 1H), 6.76 (t, J = 7.5 Hz, 1H), 6.70 (d, J = 12.3 Hz, 1H), 6.64 (d, J = 12.2 Hz, 1H), 3.83 (s, 3H). **¹³C NMR** (151 MHz, CDCl₃) δ 157.2, 137.3, 130.2, 130.1, 128.8, 128.6, 128.0, 126.9, 126.2, 125.8, 120.2, 110.6, 55.4.

(Z)-methyl(4-styrylphenyl)sulfane (12g)[191]

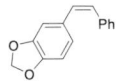

80%, white solid. **¹H NMR** (400 MHz, CDCl₃) δ 7.28-7.19 (m, 5H), 7.18 (d, *J* = 8.5 Hz, 2H), 7.10 (d, *J* = 8.4 Hz, 2H), 6.55 (q, *J* = 12.2 Hz, 2H), 2.46 (s, 3H). **¹³C NMR** (101 MHz, CDCl₃) δ 137.3, 137.2, 133.9, 130.0, 129.6, 129.3, 128.8, 128.3, 127.1, 126.1, 15.6. **MS (EI)**: m/z = 226.7 [M]⁺. **IR (ATR)**: ν = 3468, 3015, 2919, 2854, 1952, 1898, 1593, 1491, 1439, 1320, 1186, 1090, 1017, 962, 920, 869, 819, 778, 732, 698, 560, 515, 469 cm⁻¹. **HRMS (ESI)**: calc. for C₁₅H₁₄S [M]⁺: 226.0811, found 226.0811

(Z)-5-styrylbenzo[d][1,3]dioxole (12h)[192]

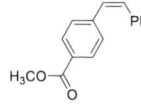

72%, white solid. **¹H NMR** (400 MHz, CDCl₃) δ 7.31 – 7.14 (m, 5H), 6.75-6.68 (m, 3H), 6.55 – 6.44 (m, 2H), 5.90 (s, 2H). **¹³C NMR** (101 MHz, CDCl₃) δ 147.3, 146.6, 137.3, 131.1, 129.8, 129.2, 128.8, 128.2, 127.0, 122.9, 108.9, 108.1, 100.9.

Methyl *(Z)*-4-styrylbenzoate (12i)[189]

82%, colorless oil. **¹H NMR** (400 MHz, CDCl₃) δ 7.90 (d, *J* = 8.4 Hz, 2H), 7.30 (d, *J* = 8.1 Hz, 2H), 7.22 (s, 5H), 6.70 (d, *J* = 12.2 Hz, 1H), 6.60 (d, *J* = 12.3 Hz, 1H), 3.88 (s, 3H). **¹³C NMR** (101 MHz, CDCl₃) δ 166.8, 142.1, 136.6, 132.2, 129.5, 129.2, 128.8, 128.6, 128.3, 127.5, 52.0.

(Z)-4-styryl-1,1'-biphenyl (12j)[190]

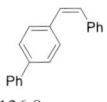

86%, white solid. **¹H NMR** (400 MHz, CDCl₃) δ 7.62 – 7.57 (m, 2H), 7.52 – 7.39 (m, 4H), 7.37 – 7.30 (m, 5H), 7.29 – 7.16 (m, 3H), 6.69 – 6.58 (m, 2H). **¹³C NMR** (101 MHz, CDCl₃) δ 140.7, 139.8, 137.3, 136.2, 130.4, 129.8, 129.3, 128.9, 128.7, 128.3, 127.3, 127.2, 126.9, 126.8.

(Z)-1-styrylnaphthalene (12k)[187]

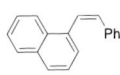

78%, yellow oil. **¹H NMR** (600 MHz, CDCl₃) δ 8.10 (d, J = 7.2 Hz, 1H), 7.89 (d, J = 7.3 Hz, 1H), 7.79 (d, J = 7.3 Hz, 1H), 7.52-7.26 (m, 5H), 7.15 – 7.03 (m, 5H), 6.86 (d, J = 12.1 Hz, 1H).**¹³C NMR** (151 MHz, CDCl₃) δ 136.7, 135.3, 133.7, 132.0, 131.6, 129.1, 128.5, 128.4, 128.0, 127.5, 127.1, 126.4, 126.0, 126.0, 125.6, 124.9.

(Z)-1-chloro-4-styrylbenzene (12l)[187]

85%, colorless oil. **¹H NMR** (600 MHz, CDCl₃) δ 7.28 – 7.14 (m, 9H), 6.63 (d, J = 12.2 Hz, 1H), 6.53 (d, J = 12.2 Hz, 1H). **¹³C NMR** (151 MHz, CDCl₃) δ 136.8, 135.2, 132.7, 130.9, 130.2, 128.9, 128.8, 128.4, 128.3, 127.3.

(Z)-1-fluoro-4-styrylbenzene (12m)[187]

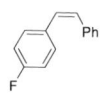

82%, colorless oil. **¹H NMR** (600 MHz, CDCl₃) δ 7.30 – 7.19 (m, 7H), 6.96 – 6.89 (m, 2H), 6.59 (dd, J = 31.0, 12.2 Hz, 2H). **¹³C NMR** (151 MHz, CDCl₃) δ 161.8 (d, J = 246.6 Hz), 137.0, 133.2 (d, J = 3.8 Hz), 130.5 (d, J = 7.9 Hz), 130.2, 129.1, 128.8, 128.3, 127.2, 115.2 (d, J = 21.4 Hz). **¹⁹F NMR** (282 MHz, CDCl₃) δ -114.60 – -114.83 (m).

(Z)-1-bromo-3-styrylbenzene (12n)[137]

75%, colorless oil. **¹H NMR** (600 MHz, CDCl₃) δ 7.42 (s, 1H), 7.34 (d, J = 7.9 Hz, 1H), 7.31 – 7.22 (m, 5H), 7.18 (d, J = 7.6 Hz, 1H), 7.09 (t, J = 7.8 Hz, 1H), 6.67 (d, J = 12.2 Hz, 1H), 6.54 (d, J = 12.2 Hz, 1H). **¹³C NMR** (151 MHz, CDCl₃) δ 139.4, 136.6, 131.8, 131.6, 130.1, 129.7, 128.8, 128.6, 128.3, 127.5, 127.4, 122.3.

(Z)-4-styrylpyridine (12o)[193]

77%, colorless oil. **¹H NMR** (600 MHz, CDCl₃) δ 8.45 (d, J = 5.5 Hz, 2H), 7.26-7.20 (m, 5H), 7.11 (d, J = 5.6 Hz, 2H), 6.79 (d, J = 12.2 Hz, 1H), 6.50 (d, J = 12.2 Hz, 1H). **¹³C NMR** (151 MHz, CDCl₃) δ 149.8, 145.0, 136.1, 134.1, 128.7, 128.4, 127.8, 127.5, 123.5.

(Z)-2-styrylthiophene (12p)[189]

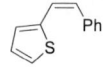

90%, yellow oil. **¹H NMR** (400 MHz, CDCl₃) δ 7.43 – 7.17 (m, 5H), 7.08 (d, J = 5.1 Hz, 1H), 6.96 (d, J = 3.5 Hz, 1H), 6.88 (dd, J = 5.0, 3.6 Hz, 1H), 6.70 (d, J = 11.9 Hz, 1H), 6.58 (d, J = 12.0 Hz, 1H). **¹³C NMR** (101 MHz, CDCl₃) δ 139.8, 137.3, 128.9, 128.8, 128.5, 128.1, 127.5, 126.39, 125.5, 123.3.

(Z)-1-methyl-5-styryl-1H-indole (12q)

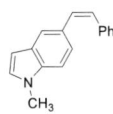

79%, yellow oil. **¹H NMR** (600 MHz, CDCl₃) δ 7.60 (s, 1H), 7.36 (d, *J* = 6.9 Hz, 2H), 7.29 – 7.16 (m, 5H), 7.04 (s, 1H), 6.79 (d, *J* = 12.2 Hz, 1H), 6.58 (d, *J* = 12.2 Hz, 1H), 6.44 (t, *J* = 2.4 Hz, 1H), 3.77 (s, 3H). **¹³C NMR** (151 MHz, CDCl₃) δ 138.0, 136.0, 131.6, 129.1, 129.0, 128.4, 128.4, 128.1, 128.0, 126.7, 122.9, 121.5, 108.9, 101.3, 32.9. **MS (EI)**: m/z = 233.2 [M]⁺. **IR (ATR)**: ν = 3466, 3012, 2926, 1953, 1883, 1697, 1615, 1488, 1447, 1412, 1378, 1332, 1243, 1156, 1079, 1023, 966, 890, 849, 799, 724, 619, 595, 536, 471 cm⁻¹. **HRMS (ESI)**: calc. for C₁₇H₁₅NNa [M + Na]⁺: 256.1097, found 256.1094

(Z)-triisopropyl(styryl)silane (12r)[194]

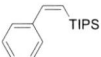

76%, colorless oil. **¹H NMR** (600 MHz, CDCl₃) δ 7.55 (dd, *J* = 15.6, 1.6 Hz, 1H), 7.34 – 7.23 (m, 5H), 5.77 (dd, *J* = 15.6, 2.4 Hz, 1H), 1.12-1.06 (m, 3H), 1.01 (d, *J* = 7.2 Hz, 18H). **¹³C NMR** (151 MHz, CDCl₃) δ 148.0, 140.7, 127.8, 127.6, 127.5, 127.3, 18.9, 12.5.

(Z)-but-1-ene-1,4-diyldibenzene (12s)[195]

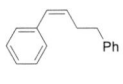

69%, light brown oil. **¹H NMR** (600 MHz, CDCl₃) δ 7.43 – 7.22 (m, 10H), 6.53 (d, J = 11.7 Hz, 1H), 5.81-5.77 (m, 1H), 2.88 – 2.82 (m, 2H), 2.79 – 2.73 (m, 2H). **¹³C NMR** (151 MHz, CDCl₃) δ 141.7, 137.6, 131.9, 129.5, 128.8, 128.6, 128.4, 128.2, 126.7, 126.0, 36.2, 30.5.

(Z)-(3-(benzyloxy)prop-1-en-1-yl)benzene (12t)[196]

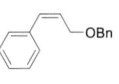

91%, colorless oil. **¹H NMR** (400 MHz, CDCl₃) δ 7.37 – 7.15 (m, 10H), 6.62 (d, *J* = 11.8 Hz, 1H), 5.95 – 5.85 (m, 1H), 4.52 (s, 2H), 4.30 (dd, *J* = 6.4, 1.6 Hz, 2H). **¹³C NMR** (101 MHz, CDCl₃) δ 138.1, 136.6, 131.8, 128.9, 128.7, 128.3, 128.2, 127.8, 127.6, 127.1, 72.5, 66.9.

(Z)-tert-butyldimethyl((3-phenylallyl)oxy)silane (12u)

81%, colorless oil. **¹H NMR** (600 MHz, CDCl₃) δ 7.36 (t, J = 7.6 Hz, 2H), 7.27 (t, J = 6.9 Hz, 1H), 7.21 (d, *J* = 7.3 Hz, 2H), 6.52 (d, *J* = 11.8 Hz, 1H), 5.88-5.84 (m, J = 11.9, 6.0 Hz, 1H), 4.48 (dd, *J* = 6.1, 1.7 Hz, 2H), 0.93 (s, 9H), 0.08 (s, 6H). **¹³C NMR** (151 MHz, CDCl₃) δ 136.87, 132.60, 129.56, 128.80, 128.18, 127.01, 60.38, 25.95, 18.34, -5.11. **MS (EI)**: m/z = 248.2 [M]⁺. **IR (ATR)**: ν = 2930, 2858, 2329, 2087, 1884, 1600, 1463, 1359, 1253, 1089, 955, 839, 774, 696 cm⁻¹. **HRMS (ESI)**: calc. for C₁₄H₂₁OSi [M-CH₃]⁺: 233.1356, found 233.1357

(Z)-tert-butyldimethyl((3-(p-tolyl)allyl)oxy)silane (12v)

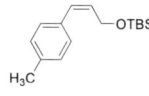

94%, colorless oil. **¹H NMR** (400 MHz, CDCl₃) δ 7.14 (d, *J* = 8.1 Hz, 2H), 7.09 (d, *J* = 8.1 Hz, 2H), 6.45 (d, *J* = 11.8 Hz, 1H), 5.81-5.75 (m, 1H), 4.45 (dd, *J* = 6.0, 1.7 Hz, 2H), 2.35 (s, 3H), 0.90 (s, 9H), 0.06 (s, 6H). **¹³C NMR** (101 MHz, CDCl₃) δ 136.7, 134.0, 131.8, 129.4, 128.8, 128.7, 60.4, 25.9, 21.2, 18.3, -5.1. **MS (EI)**: m/z = 262.2 [M]⁺. **IR (ATR)**: ν = 2931, 2889, 2858, 2162, 1689, 1610, 1513, 1466, 1389, 1361, 1254, 1088, 1008, 963, 832, 776, 671 cm⁻¹. **HRMS (ESI)**: calc. for C₁₆H₂₆OSi [M]⁺: 262.1747, found 262.1745

(Z)-tert-butyl((3-(4-chlorophenyl)allyl)oxy)dimethylsilane (12w)

73%, white solid. ^{1}H NMR (600 MHz, CDCl$_3$) δ 7.30 (d, *J* = 8.4 Hz, 2H), 7.12 (d, *J* = 8.3 Hz, 2H), 6.44 (d, *J* = 11.8 Hz, 1H), 5.86-5.82 (m, 1H), 4.40 (d, *J* = 6.1 Hz, 2H), 0.89 (s, 9H), 0.05 (s, 6H). ^{13}C NMR (151 MHz, CDCl$_3$) δ 135.2, 133.1, 132.8, 130.0, 128.5, 128.3, 60.1, 25.9, 18.3, -5.2. **MS (EI)**: m/z = 225.1 [M-tBu]$^+$. **IR (ATR)**: *v* = 2931, 2889, 2858, 2171, 1720, 1594, 1491, 1468, 1392, 1360, 1254, 1090, 1011, 962, 835, 777, 674 cm^{-1}. **HRMS (ESI)**: calc. for C$_{11}$H$_{14}$OClSi [M-tBu]$^+$: 225.0497, found 225.0497

(Z)-tert-butyldimethyl((3-(thiophen-2-yl)allyl)oxy)silane (12x)

76%, yellow oil. ^{1}H NMR (600 MHz, CDCl$_3$) δ 7.29 (d, *J* = 4.9 Hz, 1H), 7.04 – 6.98 (m, 1H), 6.93 (d, *J* = 2.8 Hz, 1H), 6.53 (d, *J* = 11.8 Hz, 1H), 5.75-5.71 (m, 1H), 4.57-4.55 (m, 2H), 0.92 (s, 9H), 0.09 (s, 6H). ^{13}C NMR (151 MHz, CDCl$_3$) δ 140.0, 131.3, 127.5, 127.1, 125.9, 120.9, 61.1, 25.9, 18.3, -5.2. **MS (EI)**: m/z = 254.3 [M]$^+$. **IR (ATR)**: *v* = 2931, 2889, 2857, 2167, 1466, 1400, 1358, 1254, 1215, 1091, 1005, 941, 834, 775, 692 cm^{-1}. **HRMS (ESI)**: calc. for C$_{13}$H$_{22}$OSSi [M]$^+$: 254.1155, found 254.1154

(Z)-4-(3-((tert-butyldimethylsilyl)oxy)prop-1-en-1-yl)pyridine (12y)

82%, colorless oil. ^{1}H NMR (400 MHz, CDCl$_3$) δ 8.53 (d, *J* = 4.6 Hz, 2H), 7.05 (d, *J* = 4.7 Hz, 2H), 6.37 (d, *J* = 11.9 Hz, 1H), 5.99 (ddd, *J* = 12.0, 6.1, 3.1 Hz, 1H), 4.39 (dd, *J* = 6.1, 1.6 Hz, 2H), 0.87 – 0.84 (m, 9H), 0.05 – -0.01 (m, 6H). ^{13}C NMR (101 MHz, CDCl$_3$) δ 149.7, 144.1, 136.7, 127.1, 123.3, 60.0, 25.8, 18.2, -5.2. **MS (EI)**: m/z = 249.9 [M]$^+$. **IR (ATR)**: *v* = 3441, 3025, 2931, 2890, 2856, 1931, 1657, 1597, 1551, 1466, 1414, 1382, 1255, 1128, 1068, 965, 877, 835, 778, 673, 636, 522 cm^{-1}. **HRMS (ESI)**: calc. for C$_{14}$H$_{24}$ONSi [M+H]$^+$: 250.1622, found 250.1621.

4.4 Hydrogenation and dehydrogenation of heterocycles with the application of manganese pincer complexes

4.4.1 General information

All reactions were carried out under an argon atmosphere using an oven-dried glassware. The dry and degassed *t*-amyl alcohol was distilled from calcium hydride under nitrogen. The dry and degassed dioxane was distilled from solvona and benzophenone under nitrogen. Toluene was obtained from MBRAUN Solvent Purification System. All other chemicals were used as purchased without further purification. ^{1}H, ^{13}C and ^{19}F spectra were recorded in CDCl$_3$ using Varian VNMR 600 MHz or Inova 400 MHz spectrometer. The signals were referenced to residual chloroform (7.26 ppm, ^{1}H, 77.00 ppm, ^{13}C). Chemical shifts are reported in ppm, multiplicities are indicated by s (singlet), d (doublet), t (triplet), dd (doublet of doublets), dt (doublet of triplets), td (triplet of doublets), br (broad signal) and m (multiplet). Analytical thin-layer chromatography (TLC) was performed using silica gel 60 pre-coated aluminium plates (Macherey-Nagel 0.20 mm thickness) with a fluorescent indicator UV254. Visualization was performed with standard phosphomolybdic acid stain (10g in 100 mL EtOH) or UV light. Analysis by gas chromatography (GC) was done using a CP-Sil-8-CB column (30 m, d = 0.25 mm) with FID detector and H_2 as carrier gas. Substrates **15k-15o** were synthesized using procedure described in our previous report.[197] Other substrates were purchased and used without further purification.

4.4.2 General procedure and reaction analysis

General procedure for the hydrogenation of indoles

$$\text{15} \xrightarrow[\text{toluene 0.25 M}]{\text{Mn-6, KO}^t\text{Bu}, \text{H}_2 \text{ (50 bar)}, 100\ °\text{C}, 36\ \text{h}} \text{16}$$

In an argon filled glovebox a 15 mL glass vial was charged with the corresponding indole **15** (0.25 mmol), **Mn-6** (2-5 mol%), KO^{t}Bu (5-12.5 mol%) and 1 mL of degassed toluene. The vial was sealed with a cap with a septum and was transferred into a stainless steel autoclave and a hole was made with a needle to allow an access of the gases. The autoclave was carefully flushed three times with nitrogen and then hydrogen gas. After adjusting the final hydrogen pressure to 50 bar, the autoclave was heated to 100-130 °C for 36 h with stirring. After cooling down the autoclave to room temperature the residual H_2 was carefully released and the mixture was analysed by TLC. Next, the reaction mixtures were purified by column chromatography on silica gel to obtain the pure indolines **16**.

General procedure for the hydrogenation of the other *N*-containing heterocycles

$$\text{17 (X=N, O)} \xrightarrow[\text{toluene 0.25 M}]{\text{Mn-6, KO}^t\text{Bu}, \text{H}_2 \text{ (50 bar)}, 120\ °\text{C}, 24\ \text{h}} \text{18}$$

In an argon filled glovebox a 15 mL glass vial was charged with the corresponding heterocycle **17** (0.25 mmol), **Mn-6** (1-3 mol%), KO^{t}Bu (2.5-7.5 mol%) and 1 mL of degassed toluene. The vial was sealed with a cap with a septum and was transferred into a stainless steel autoclave and a hole was made with a needle to allow an access of the gases. The autoclave was carefully flushed three times with nitrogen and then hydrogen gas. After adjusting the final hydrogen pressure to 50 bar, the autoclave was heated to 120-140 °C for 24 h with stirring. After cooling down the autoclave to room temperature the residual H_2 was carefully released and the mixture was analysed by TLC. Next, the reaction mixtures were purified by column chromatography on silica gel to obtain the pure reduced heterocycles **18**.

Mn-1 catalysed scaled up hydrogenation of indoles

In an argon filled glovebox a 15 mL glass vial was charged with the corresponding indole **15** (2.5 mmol), **Mn-6** (2 mol%), KO^{t}Bu (5 mol%) and 5 mL of degassed toluene. The vial was sealed with a cap with a septum and transferred into a stainless steel autoclave and a hole was made with a needle to allow an access of the gases. The autoclave was carefully flushed three times with nitrogen and then hydrogen gas. After adjusting the final hydrogen pressure to 50 bar, the autoclave was heated to 100 °C (130 °C for the substrate **15d**) for 36 h with stirring. After cooling down to room temperature the residual H$_2$ was vented carefully and was analysed by TLC. Next, the reaction mixture was purified by column chromatography on silica gel to obtain pure indolines **16**.

General procedure for the dehydrogenation of *N*-containing heterocycles

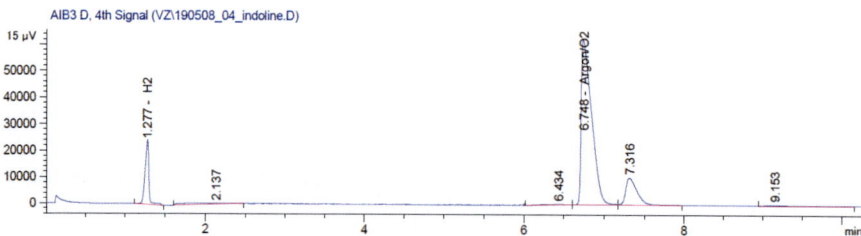

Dehydrogenation reaction was performed using reduced heterocycles **16a, 16j, 18a, 18b, 18c**. In an argon filled glovebox a 38 mL pressure vessel was charged with the reduced heterocycle **16** or **18** (0.25 mmol), **Mn-1** (1-10 mol%), KO^{t}Bu (2.5-25 mol%) and 1 mL of toluene. The pressure vessel was closed tight with a cap and stirred at 120-160 °C for 24 h. After cooling down to a room temperature the cap was slowly opened, and the mixture was analysed by TLC. Next, the reaction mixture was purified by column chromatography on silica gel to obtain pure dehydrogenated heterocycles **15, 17**.

Figure 14. Detection of H$_2$ gas from the catalytic dehydrogenation of indoline was performed using a Agilent 7890B Gas Analyzer. GC parameters: injection temperature = 60 °C, column temperature = 60 °C, detector temperature = 250 ° C. The carrier gas - N$_2$.

Spectral data
Indoline, 16a[198]

83%, brown liquid. **¹H NMR** (600 MHz, CDCl₃) δ 7.14 (d, *J* = 7.3 Hz, 1H), 7.03 (t, *J* = 7.5 Hz, 1H), 6.72 (t, *J* = 7.4 Hz, 1H), 6.66 (d, *J* = 7.7 Hz, 1H), 3.65 (br, 1H), 3.57 (t, *J* = 7.2 Hz, 2H), 3.05 (t, *J* = 8.1 Hz, 2H). **¹³C NMR** (151 MHz, CDCl₃) δ 151.5, 129.3, 127.2, 124.6, 118.7, 109.5, 47.4, 29.8.

7-Methylindoline, 16b[199]

91%, brown oil. **¹H NMR** (400 MHz, CDCl₃) δ 7.00 (d, *J* = 7.3 Hz, 1H), 6.87 (d, *J* = 7.5 Hz, 1H), 6.67 (t, *J* = 7.4 Hz, 1H), 3.58 (t, *J* = 8.4 Hz, 12H), 3.06 (t, *J* = 8.4 Hz, 2H), 2.15 (s, 3H). **¹³C NMR** (101 MHz, CDCl₃) δ 150.0, 128.6, 128.1, 122.1, 119.0, 118.9, 47.1, 30.1, 16.8.

6-Methylindoline, 16c[200]

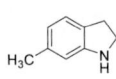
69%, brown oil. **¹H NMR** (400 MHz, CDCl₃) δ 6.98 (d, *J* = 7.4 Hz, 1H), 6.51 (d, *J* = 7.4 Hz, 1H), 6.48 (s, 1H), 3.53 (t, *J* = 8.0 Hz, 2H), 2.97 (t, *J* = 8.1 Hz, 2H), 2.25 (s, 3H). **¹³C NMR** (101 MHz, CDCl₃) δ 151.7, 137.0, 126.4, 124.2, 119.4, 110.4, 47.6, 29.5, 21.4.

5-Methoxyindoline, 16d[198]

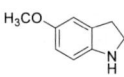
85%, brown oil. **¹H NMR** (400 MHz, CDCl₃) δ 6.76 (s, 1H), 6.60 (d, *J* = 1.4 Hz, 2H), 3.75 (s, 3H), 3.54 (t, *J* = 8.3 Hz, 2H), 3.29 (br, 1H), 3.01 (t, *J* = 8.3 Hz, 2H). **¹³C NMR** (101 MHz, CDCl₃) δ 153.5, 145.2, 131.2, 112.1, 111.5, 110.1, 56.0, 47.8, 30.4.

Methyl indoline-5-carboxylate, 16e[201]

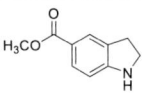
97%, light brown solid. **¹H NMR** (400 MHz, CDCl₃) δ 7.80 – 7.64 (m, 2H), 6.51 (d, *J* = 8.6 Hz, 1H), 4.13 (br, 1H), 3.82 (s, 3H), 3.62 (t, *J* = 8.6 Hz, 2H), 3.03 (t, *J* = 8.5 Hz, 2H). **¹³C NMR** (101 MHz, CDCl₃) δ 167.4, 155.9, 130.7, 128.6, 126.1, 119.5, 51.5, 47.2, 28.8.

5-Bromoindoline, 16f[202]

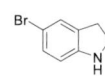
99%, brown oil. **¹H NMR** (600 MHz, CDCl₃) δ 7.19 (s, 1H), 7.09 (d, *J* = 8.2 Hz, 1H), 6.50 (d, *J* = 8.2 Hz, 1H), 3.56 (t, *J* = 8.5 Hz, 2H), 3.01 (t, *J* = 8.4 Hz, 2H). **¹³C NMR** (151 MHz, CDCl₃) δ 150.7, 131.7, 129.8, 127.6, 110.5, 110.1, 47.6, 29.7.

4-Bromoindoline, 16g[203]

86%, brown oil. **¹H NMR** (600 MHz, CDCl₃) δ 6.87 (t, *J* = 7.7 Hz, 1H), 6.81 (d, *J* = 8.0 Hz, 1H), 6.52 (d, *J* = 7.5 Hz, 1H), 3.59 (t, *J* = 8.4 Hz, 2H), 3.05 (t, *J* = 8.4 Hz, 2H). **¹³C NMR** (151 MHz, CDCl₃) δ 152.6, 129.8, 128.9, 121.3, 119.7, 107.8, 46.4, 31.1.

6-Fluoroindoline, 16h[204]

99%, brown oil. **¹H NMR** (600 MHz, CDCl₃) δ 7.00 – 6.96 (m, 1H), 6.38 – 6.29 (m, 2H), 3.59 (t, *J* = 8.3 Hz, 2H), 2.97 (t, *J* = 8.2 Hz, 2H). **¹³C NMR** (151 MHz, CDCl₃) δ 163.0 (d, *J*

= 240.6 Hz), 153.0 (d, J = 11.7 Hz), 124.8 (d, J = 10.3 Hz), 124.5 (d, J = 1.8 Hz), 104.4 (d, J = 22.6 Hz), 97.0 (d, J = 26.2 Hz), 48.1, 28.9. ^{19}**F NMR** (564 MHz, CDCl$_3$) δ -116.6 (dd, J = 15.2, 9.3 Hz).

5-Fluoroindoline, 16i[198]

86%, brown oil. ^{1}**H NMR** (600 MHz, CDCl$_3$) δ 6.84 (d, J = 8.4 Hz, 1H), 6.71 (td, J = 8.9, 2.5 Hz, 1H), 6.55 (dd, J = 8.4, 4.3 Hz, 1H), 3.63 (br, 1H), 3.57 (t, J = 8.4 Hz, 2H), 3.02 (t, J = 8.3 Hz, 2H). ^{13}**C NMR** (151 MHz, CDCl$_3$) δ 157.0 (d, J = 235.0 Hz), 147.3, 131.2 (d, J = 8.3 Hz), 113.1 (d, J = 23.4 Hz), 112.0 (d, J = 23.8 Hz), 109.6 (d, J = 8.4 Hz), 47.9, 30.2. ^{19}**F NMR** (564 MHz, CDCl$_3$) δ -126.56.

3-Methylindoline, 16j[199]

78%, brown oil. ^{1}**H NMR** (400 MHz, CDCl$_3$) δ 7.07 (d, J = 7.2 Hz, 1H), 7.01 (t, J = 7.5 Hz, 1H), 6.72 (t, J = 7.3 Hz, 1H), 6.63 (d, J = 7.7 Hz, 1H), 3.69 (t, J = 8.6 Hz, 1H), 3.41 – 3.29 (m, 1H), 3.10 (t, J = 8.5 Hz, 1H), 1.31 (d, J = 6.8 Hz, 3H). ^{13}**C NMR** (101 MHz, CDCl$_3$) δ 151.2, 134.3, 127.2, 123.3, 118.7, 109.5, 55.4, 36.6, 18.6.

3-Benzylindoline, 16k[205]

78%, brown oil. ^{1}**H NMR** (400 MHz, CDCl$_3$) δ 7.34 – 7.18 (m, 5H), 7.04 (t, J = 7.6 Hz, 1H), 6.96 (d, J = 7.3 Hz, 1H), 6.73 – 6.62 (m, 2H), 3.66 – 3.49 (m, 2H), 3.32 – 3.24 (m, 1H), 3.11 (dd, J = 13.7, 5.8 Hz, 1H), 2.82 (dd, J = 13.7, 8.9 Hz, 1H). ^{13}**C NMR** (101 MHz, CDCl$_3$) δ 151.3, 140.1, 132.4, 129.0, 128.4, 127.6, 126.1, 124.1, 118.6, 109.7, 52.9, 43.6, 40.4.

3-Phenethylindoline, 16l

65%, brown oil. ^{1}**H NMR** (600 MHz, CDCl$_3$) δ 7.31-7.29 (m, 2H), 7.23-7.19 (m, 3H), 7.12 (d, J = 7.3 Hz, 1H), 7.05 (t, J = 7.5 Hz, 1H), 6.74 (t, J = 7.3 Hz, 1H), 6.67 (d, J = 7.7 Hz, 1H), 3.72 (t, J = 8.4 Hz, 1H), 3.37 – 3.22 (m, 2H), 2.73 (t, J = 8.0 Hz, 2H), 2.23 – 2.12 (m, 1H), 1.93 – 1.84 (m, 1H). ^{13}**C NMR** (151 MHz, CDCl$_3$) δ 151.2, 142.0, 132.9, 128.4, 128.4, 127.5, 125.9, 123.9, 118.7, 109.7, 53.4, 41.7, 35.9, 33.7. **MS (EI)**: m/z = 223.2 [M]$^+$. **IR (ATR)**: ν =3382, 3028, 2924, 2853, 1603, 1484, 1461, 1318, 1244, 1151, 1092, 1024, 922, 846, 740, 700 cm^{-1}. **HRMS (ESI)**: calc. for C$_{16}$H$_{18}$N [M + H]$^+$: 224.1433, found 224.1433.

3-(4-Methoxybenzyl)indoline, 16m

72%, brown oil. ^{1}**H NMR** (600 MHz, CDCl$_3$) δ 7.15 (d, J = 8.6 Hz, 2H), 7.06 (t, J = 7.6 Hz, 1H), 6.98 (d, J = 7.3 Hz, 1H), 6.90 – 6.86 (m, 2H), 6.72 (td, J = 7.4, 0.7 Hz, 1H), 6.67 (d, J = 7.7 Hz, 1H), 3.83 (s, 3H), 3.64 (br, 1H), 3.61 – 3.52 (m, 2H), 3.32 – 3.25 (m, 1H), 3.09 – 3.04 (m, 1H), 2.82 – 2.75 (m, 1H). ^{13}**C NMR** (151 MHz, CDCl$_3$) δ 158.0, 151.4, 132.5, 132.2, 129.9, 127.6, 124.2, 118.6, 113.8, 109.7, 55.3, 52.9, 43.8, 39.5. **MS (EI)**: m/z = 239.2 [M]$^+$. **IR (ATR)**: ν = 3379, 3035, 2926, 2839, 1606, 1507, 1462, 1303, 1241, 1175, 1107, 1031, 911, 819, 739 cm^{-1}. **HRMS (ESI)**: calc. for C$_{16}$H$_{18}$ON [M + H]$^+$: 240.1382, found 240.1378.

3-(2-Chlorobenzyl)indoline, 16n

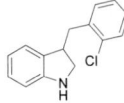

76%, brown oil. **¹H NMR** (600 MHz, CDCl₃) δ 7.29 (d, *J* = 8.1 Hz, 2H), 7.14 (d, *J* = 8.2 Hz, 2H), 7.06 (t, *J* = 7.6 Hz, 1H), 6.94 (d, *J* = 7.3 Hz, 1H), 6.71 (t, *J* = 7.4 Hz, 1H), 6.67 (d, *J* = 7.7 Hz, 1H), 3.61 – 3.50 (m, 2H), 3.26 (s, 1H), 3.06 (dd, *J* = 13.8, 5.4 Hz, 1H), 2.81 (dd, *J* = 13.8, 8.1 Hz, 1H). **¹³C NMR** (151 MHz, CDCl₃) δ 151.3, 138.5, 132.0, 131.9, 130.4, 128.5, 127.8, 124.1, 118.6, 109.8, 52.8, 43.5, 39.7. **MS (EI):** m/z = 243.1 [M]⁺. **IR (ATR):** *v* =3381, 3036, 2927, 2851, 2326, 2073, 1900, 1779, 1604, 1485, 1404, 1321, 1244, 1162, 1091, 1015, 927, 840, 800, 744, 661 cm⁻¹. **HRMS (ESI):** calc. for C₁₅H₁₅NCl [M + H]⁺: 244.0887, found 244.0885.

3-(Cyclopropylmethyl)indoline, 16o

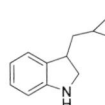

69%, brown oil. **¹H NMR** (600 MHz, CDCl₃) δ 7.11 (d, *J* = 7.3 Hz, 1H), 7.03 (t, *J* = 7.6 Hz, 1H), 6.72 (t, *J* = 7.4 Hz, 1H), 6.66 (d, *J* = 7.7 Hz, 1H), 3.75 (t, *J* = 8.5 Hz, 1H), 3.63 (br, 1H), 3.41-3.36 (m, 1H), 3.31 (t, *J* = 8.1 Hz, 1H), 1.74 – 1.66 (m, 1H), 1.53 – 1.46 (m, 1H), 0.83 – 0.73 (m, 1H), 0.53 – 0.42 (m, 2H), 0.16 – 0.04 (m, 2H). **¹³C NMR** (151 MHz, CDCl₃) δ 151.3, 133.2, 127.3, 123.9, 118.6, 109.6, 53.5, 42.7, 38.9, 9.3, 4.8, 4.5. **MS (EI):** m/z = 173.1 [M]⁺. **IR (ATR):** *v* = 3378, 3072, 3000, 2911, 2847, 2325, 2045, 1918, 1771, 1605, 1470, 1317, 1240, 1142, 1100, 1017, 926, 824, 740 cm⁻¹. **HRMS (ESI):** calc. for C₁₂H₁₆N [M + H]⁺: 174.1277, found 174.1277.

1,2,3,4-Tetrahydroquinoxaline, 18a[202]

99%, light brown solid. **¹H NMR** (500 MHz, CDCl₃) δ 6.56 (d, *J* = 42.4 Hz, 4H), 3.66 (br, 2H), 3.44 (s, 4H). **¹³C NMR** (126 MHz, CDCl₃) δ 133.7, 118.8, 114.7, 41.4.

2-Methyl-1,2,3,4-tetrahydroquinoxaline, 18b[202]

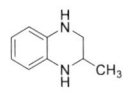

99%, light brown solid. **¹H NMR** (400 MHz, CDCl₃) δ 6.63-6.59 (m, 2H), 6.55-6.51 (m, 2H), 3.67 (br, 1H), 3.61 – 3.48 (m, 2H), 3.35 (dd, *J* = 10.7, 2.6 Hz, 1H), 3.07 (dd, *J* = 10.5, 8.3 Hz, 1H), 1.22 (d, *J* = 6.2 Hz, 3H). **¹³C NMR** (101 MHz, CDCl₃) δ 133.6, 133.2, 118.7, 114.5, 114.4, 48.3, 45.7, 19.9.

2-Phenyl-1,2,3,4-tetrahydroquinoxaline, 18c[206]

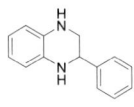

99%, light orange solid. **¹H NMR** (400 MHz, CDCl₃) δ 7.49 – 7.30 (m, 5H), 6.78 – 6.47 (m, 4H), 4.58 – 4.40 (m, 1H), 3.86 (br, 2H), 3.55 – 3.43 (m, 1H), 3.40 – 3.30 (m, 1H). **¹³C NMR** (101 MHz, CDCl₃) δ 141.8, 134.2, 132.7, 128.7, 127.9, 127.0, 119.0, 118.8, 114.8, 114.5, 54.7, 49.2.

3-Phenyl-3,4-dihydro-2H-benzo[b][1,4]oxazine, 18d[207]

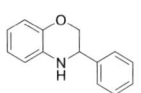

99%, light brown oil. **¹H NMR** (400 MHz, CDCl₃) δ 7.31-7.25 (m, 5H), 6.82 – 6.67 (m, 2H), 6.63-6.57 (m, 2H), 4.40 (d, *J* = 8.3 Hz, 1H), 4.19 (d, *J* = 12.0 Hz, 1H), 3.90 (t, *J* = 9.4 Hz, 2H). **¹³C NMR** (101 MHz, CDCl₃) δ 143.6, 139.2, 133.9, 128.9, 128.4, 127.2, 121.5, 119.0, 116.6, 115.4, 71.0, 54.3.

Indole, 15a[208]

89%, colorless solid. **¹H NMR** (600 MHz, CDCl₃) δ 8.12 (br, 1H), 7.67 (d, *J* = 7.9 Hz, 1H), 7.41 (d, *J* = 8.1 Hz, 1H), 7.24 – 7.19 (m, 2H), 7.14 (t, *J* = 7.5 Hz, 1H), 6.60 – 6.55 (m, 1H). **¹³C NMR** (151 MHz, CDCl₃) δ 135.8, 127.8, 124.1, 122.0, 120.7, 119.8, 111.0, 102.6.

3-Methyl-1H-indole, 15j[208]

92%, white solid. **¹H NMR** (600 MHz, CDCl₃) δ 7.84 (br, 1H), 7.61 (d, *J* = 7.9 Hz, 1H), 7.36 (d, *J* = 8.1 Hz, 1H), 7.21 (t, *J* = 7.6 Hz, 1H), 7.15 (t, *J* = 7.4 Hz, 1H), 6.97 (s, 1H), 2.36 (d, *J* = 1.2 Hz, 3H). **¹³C NMR** (151 MHz, CDCl₃) δ 136.2, 128.3, 121.9, 121.6, 119.1, 118.8, 111.7, 110.9, 9.7.

Quinoxaline, 17a[209]

95%, white solid. **¹H NMR** (500 MHz, CDCl₃) δ 8.87 (s, 2H), 8.14 (dd, *J* = 6.3, 3.4 Hz, 2H), 7.81 (dd, *J* = 6.4, 3.4 Hz, 2H). **¹³C NMR** (126 MHz, CDCl₃) δ 145.0, 143.1, 130.1, 129.5.

2-Methylquinoxaline, 17b[210]

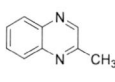
73%, dark orange solid. **¹H NMR** (500 MHz, CDCl₃) δ 8.76 (d, *J* = 2.7 Hz, 1H), 8.08 (d, *J* = 8.1 Hz, 1H), 8.03 (d, *J* = 8.1 Hz, 1H), 7.79 – 7.68 (m, 2H), 2.79 (d, *J* = 2.7 Hz, 3H). **¹³C NMR** (126 MHz, CDCl₃) δ 153.8, 146.0, 142.1, 141.0, 130.0, 129.2, 128.9, 128.7, 22.6.

2-Phenylquinoxaline, 17c[210]

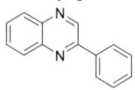
97%, white solid. **¹H NMR** (500 MHz, CDCl₃) δ 9.36 (s, 1H), 8.23 (d, *J* = 7.4 Hz, 2H), 8.19 (d, *J* = 8.2 Hz, 1H), 8.15 (d, *J* = 8.1 Hz, 1H), 7.80 (dt, *J* = 20.0, 7.0 Hz, 2H), 7.60 (t, *J* = 7.4 Hz, 2H), 7.55 (t, *J* = 7.2 Hz, 1H). **¹³C NMR** (126 MHz, CDCl₃) δ 151.9, 143.4, 142.3, 141.6, 136.8, 130.3, 130.2, 129.7, 129.6, 129.2, 129.2, 127.6.

4.5 Manganese-catalysed hydrogenation of nitroarenes

4.5.1 General information

All reactions were carried out under an argon atmosphere using oven-dried glassware. The dry and degassed *t*-amyl alcohol were distilled from calcium hydride under nitrogen. DMF, toluene and DCM were obtained from MBRAUN Solvent Purification System. The dry and degassed dioxane was distilled from sodium benzophenone under nitrogen. All other chemicals were used as purchased without further purification. ¹H, ¹³C and ¹⁹F spectra were recorded in CDCl₃ using Varian VNMR 600 MHz, Ascend Bruker 600 MHz, Ascend Bruker 400 MHz and Inova 400 MHz spectrometer. The signals were referenced to residual chloroform (7.26 ppm, ¹H, 77.00 ppm, ¹³C). Chemical shifts are reported in ppm, multiplicities are indicated by s (singlet), d (doublet), t (triplet), q (quartet), dd (doublet of doublets), td (triplet of doublets), tt (triplet of triplets) and m (multiplet). Analytical thin-layer chromatography (TLC) was performed using silica gel

60 pre-coated aluminium plates (Macherey-Nagel 0.20 mm thickness) with a fluorescent indicator UV254. Visualization was performed with standard phosphomolybdic acid stain (10g in 100 mL EtOH) or UV light. Analysis by gas chromatography (GC) was done using a CP-Sil-8-CB column (30 m, d = 0.25 mm) with FID detector and H_2 as carrier gas.

4.5.2 Experimental procedure and characterizations of the products

Synthesis of starting materials

All nitroarenes, if not otherwise noted, were purchased and used without further purification.

Compound **19n** was synthesized by alkylation of the corresponding phenol. Analytical data of the product is an agreement with the reported in the literature.[211] **^{1}H NMR** (600 MHz, CDCl$_3$) δ 8.25 – 8.17 (m, 2H), 7.47 – 7.32 (m, 5H), 7.07 – 6.99 (m, 2H), 5.16 (s, 2H). Compound **19r** was synthesized following the procedure described by Dahabiyeh, L.A. *et al.* Analytical data of the product is an agreement with the reported in the literature. [212] **^{1}H NMR** (600 MHz, CDCl$_3$) δ 8.30 (d, J = 8.7 Hz, 2H), 7.97 (d, J = 8.7 Hz, 2H), 7.44 – 7.32 (m, 5H), 6.51 (s, 1H), 4.69 (d, J = 5.6 Hz, 2H). Compound **19q** was synthesized by esterification of the corresponding acid. Analytical data of the product is an agreement with the reported in the literature.[213] **^{1}H NMR** (600 MHz, CDCl$_3$) δ 8.20 – 8.17 (m, 2H), 7.46-7.45 (m, 2H), 4.17 (q, J = 7.1 Hz, 2H), 3.72 (s, 2H), 1.26 (t, J = 7.1 Hz, 3H). Compound **19u** was synthesized following the procedure described by Uriac P. *et al.* Analytical data of the product is an agreement with the reported in the literature.[214] **^{1}H NMR** (600 MHz, CDCl$_3$) δ 8.03 (dd, J = 7.7, 1.3 Hz, 1H), 7.87 – 7.82 (m, 1H), 7.70 (td, J = 7.7, 1.4 Hz, 1H), 7.66 (td, J = 7.6, 1.1 Hz, 1H), 7.28 – 7.21 (m, 5H), 5.74 (t, J = 5.7 Hz, 1H), 4.35 (d, J = 6.3 Hz, 2H).

General procedure for the hydrogenation of nitroarenes

$$\text{Ar}-\overset{+}{\underset{O^-}{N}}\overset{O}{} + H_2 \xrightarrow[\text{toluene 0.25 M}]{\textbf{Mn-6}, K_2CO_3, 130\,°C, 24\,h} \text{Ar}-NH_2$$

19   (80 bar)   **20**

In an argon filled glovebox a 15 mL glass vial was charged with the corresponding nitroarene **19** (0.25 mmol), **Mn-6** (5-10 mol%), K$_2$CO$_3$ (12.5-25 mol%) and 1 mL of degassed toluene. The vial was sealed with a cap with a septum and was transferred into a stainless steel autoclave and a hole was made with a needle to allow an access of the gases. The autoclave was carefully flushed three times with nitrogen and then hydrogen gas. After adjusting the final hydrogen pressure to 80 bar, the autoclave was heated to 130 °C for 24 h with stirring. After cooling down the autoclave to room temperature the residual H$_2$ was carefully released and the mixture was analysed by TLC. Next, the reaction mixtures were purified by column chromatography on silica gel to obtain the pure anilines **20**.

Mn-6 catalysed hydrogenation of 1-iodo-4-nitrobenzene (scaled up experiment)

In an argon filled glovebox a 15 mL glass vial was charged with 1-iodo-4-nitrobenzene **19j** (4.02 mmol, 1000 mg), **Mn-6** (5 mol%, 127 mg), K_2CO_3 (12.5 mol%, 69.4 mg) and 4 mL of degassed toluene. The vial was sealed with a cap with a septum and was transferred into a stainless steel autoclave and a hole was made with a needle to allow an access of the gases. The autoclave was carefully flushed three times with nitrogen and then hydrogen gas. After adjusting the final hydrogen pressure to 80 bar, the autoclave was heated to 130 °C for 24 h with stirring. After cooling down the autoclave to room temperature the residual H_2 was carefully released and the mixture was analysed by TLC. Next, the reaction mixtures were purified by column chromatography on silica gel to obtain the pure 4-iodoaniline **20j** in 78% yield.

Synthesis of (2,4-dimethylphenyl)(2-nitrophenyl)sulfane (**19w**) was performed following the procedure described by M. M.Rathore[215].

(2,4-dimethylphenyl)(2-nitrophenyl)sulfane (19w): ^{1}H NMR (600 MHz, CDCl$_3$) δ 8.25 (dd, J = 8.3, 1.3 Hz, 1H), 7.47 (d, J = 7.8 Hz, 1H), 7.32-7.29 (m, 1H), 7.23 – 7.15 (m, 2H), 7.11 (d, J = 7.7 Hz, 1H), 6.70 (dd, J = 8.2, 1.1 Hz, 1H), 2.39 (s, 3H), 2.30 (s, 3H). ^{13}C NMR (151 MHz, CDCl$_3$) δ 144.9, 143.1, 141.0, 139.3, 137.2, 133.5, 132.2, 128.4, 127.3, 126.3, 126.0, 124.5, 21.3, 20.4. Analytical data of the product is an agreement with the reported in the literature[215].

2-((2,4-dimethylphenyl)thio)aniline (20w): In an argon filled glovebox a 15 mL glass vial was charged with (2,4-dimethylphenyl)(2-nitrophenyl)sulfane (**19w**) (0.77 mmol, 200 mg), **Mn-6** (5 mol%, 24.4 mg), K_2CO_3 (12.5 mol%, 13,3 mg) and 1 mL of degassed toluene. The vial was sealed with a cap with a septum and was transferred into a stainless steel autoclave and a hole was made with a needle to allow an access of the gases. The autoclave was carefully flushed three times with nitrogen and then hydrogen gas. After adjusting the final hydrogen pressure to 80 bar, the autoclave was heated to 130 °C for 24 h with stirring. After cooling down the autoclave to room temperature the residual H_2 was carefully released and the mixture was analysed by TLC. Next, the reaction mixtures were purified by column chromatography on silica gel to obtain the 2-((2,4-dimethylphenyl)thio)aniline **20w** in 74% yield. ^{1}H NMR (600 MHz, CDCl$_3$) δ 7.40 (dd, J = 7.7, 1.5 Hz, 1H), 7.28 – 7.23 (m, 1H), 7.04 (s, 1H), 6.89 (d, J = 8.0 Hz, 1H), 6.82 (dd, J = 8.0, 1.1 Hz, 1H), 6.79 (td, J = 7.5, 1.3 Hz, 1H), 6.75 (d, J = 8.0 Hz, 1H), 4.25 (s, 2H), 2.44 (s, 3H), 2.31 (s, 3H). ^{13}C NMR (151 MHz, CDCl$_3$) δ 148.4, 136.6, 135.8, 135.4, 131.8, 131.2, 130.5, 127.4, 126.6, 118.9, 115.3, 115.2, 20.8, 20.1. Analytical data of the product is an agreement with the reported in the literature[215].

General procedure for the hydrogenation of possible intermediates

In an argon filled glovebox a 15 mL glass vial was charged with the corresponding intermediate **21** (0.25 mmol), **22** (0.125 mmol) or **23** (0.125 mmol), **Mn-6** (5 mol%), K_2CO_3 (12.5 mol%) and 1 mL of degassed toluene. The vial was sealed with a cap with a septum and was transferred into a stainless steel autoclave and a hole was made with a needle to allow an access of the gases. The autoclave was carefully flushed three times with nitrogen and then hydrogen gas. After adjusting the final hydrogen pressure to 80 bar, the autoclave was heated to 130 °C for 24 h with stirring. After cooling down the autoclave to room temperature the residual H_2 was carefully released and the mixture was analysed by GC. Yields of aniline were determined by the GC analysis using dodecane as internal standard.

Spectral data

p-Toluidine, 20b[216]

78%, dark brown solid. **^{1}H NMR** (600 MHz, CDCl$_3$) δ 6.89 (d, J = 8.0 Hz, 2H), 6.53 (d, J = 8.3 Hz, 2H), 3.35 (s, 2H), 2.16 (s, 3H). **^{13}C NMR** (151 MHz, CDCl$_3$) δ 143.8, 129.8, 127.8, 115.3, 20.5.

4-(Tert-butyl)aniline, 20c[217]

97%, dark brown solid. **^{1}H NMR** (600 MHz, CDCl$_3$) δ 7.23 – 7.19 (m, 2H), 6.70 – 6.64 (m, 2H), 3.68 (s, 2H), 1.30 (s, 9H).**^{13}C NMR** (151 MHz, CDCl$_3$) δ 143.5, 141.6, 126.1, 115.1, 33.9, 31.5.

2-Ethylaniline, 20d[218]

73%, brown oil. **^{1}H NMR** (600 MHz, CDCl$_3$) δ 7.08 (d, J = 7.5 Hz, 1H), 7.05 (td, J = 7.6, 1.4 Hz, 1H), 6.76 (td, J = 7.4, 1.0 Hz, 1H), 6.72 – 6.66 (m, 1H), 3.62 (s, 2H), 2.53 (q, J = 7.5 Hz, 2H), 1.26 (t, J = 7.6 Hz, 3H). **^{13}C NMR** (151 MHz, CDCl$_3$) δ 144.0, 128.3, 128.1, 126.8, 118.8, 115.4, 24.0, 13.0.

2,6-Dimethylaniline, 20e[216]

89%, brown oil. **^{1}H NMR** (400 MHz, CDCl$_3$) δ 6.87 (d, J = 7.4 Hz, 2H), 6.57 (t, J = 7.4 Hz, 1H), 3.50 (s, 2H), 2.11 (s, 6H).**^{13}C NMR** (101 MHz, CDCl$_3$) δ 142.7, 128.2, 121.7, 118.0, 17.6.

4-Fluoroaniline, 20f[217]

83%, dark brown solid. **^{1}H NMR** (600 MHz, CDCl$_3$) δ 6.89 – 6.82 (m, 2H), 6.65 – 6.59 (m, 2H), 3.54 (s, 2H). **^{13}C NMR** (151 MHz, CDCl$_3$) δ 156.4 (d, J = 235.7 Hz), 142.4 (d, J = 1.5 Hz), 116.1 (d, J = 7.6 Hz), 115.7 (d, J = 22.8 Hz). **^{19}F NMR** (564 MHz, CDCl$_3$) δ -126.85 (tt, J = 8.7, 4.5 Hz).

4-Chloroaniline, 20g[216]

94%, grey solid. **¹H NMR** (600 MHz, CDCl₃) δ 7.14 – 7.06 (m, 2H), 6.63 – 6.59 (m, 2H), 3.59 (s, 2H). **¹³C NMR** (151 MHz, CDCl₃) δ 144.9, 129.1, 123.2, 116.3.

4-Bromoaniline, 20h[216]

93%, dark brown solid. **¹H NMR** (600 MHz, CDCl₃) δ 7.31 – 7.22 (m, 2H), 6.65 – 6.48 (m, 2H), 3.68 (s, 2H). **¹³C NMR** (151 MHz, CDCl₃) δ 145.5, 132.0, 116.7, 110.2.

4-Iodoaniline, 20i[219]

86%, dark brown solid. **¹H NMR** (600 MHz, CDCl₃) δ 7.43 – 7.38 (m, 2H), 6.50 – 6.44 (m, 2H), 3.68 (s, 2H). **¹³C NMR** (151 MHz, CDCl3) δ 146.1, 137.9, 117.3, 79.4.

3-(Trifluoromethyl)aniline, 20j[220]

65%, yellow oil. **¹H NMR** (600 MHz, CDCl₃) δ 7.27 (t, J = 7.9 Hz, 1H), 7.02 (d, J = 7.7 Hz, 1H), 6.92 (s, 1H), 6.84 (dd, J = 8.0, 1.8 Hz, 1H), 3.85 (s, 2H). **¹³C NMR** (151 MHz, CDCl₃) δ 146.7, 131.6 (q, J = 31.8 Hz), 129.7, 124.2 (q, J = 272.2 Hz), 118.0, 115.0 (q, J = 4.0 Hz), 111.3 (q, J = 3.9 Hz). **¹⁹F NMR** (565 MHz, CDCl₃) δ -62.9.

(E)-4-Styrylaniline, 20k[221]

94%, beige solid. **¹H NMR** (400 MHz, CDCl₃) δ 7.47 (d, J = 7.7 Hz, 2H), 7.35-7.31 (m, 4H), 7.21 (t, J = 7.4 Hz, 1H), 7.03 (d, J = 16.3 Hz, 1H), 6.92 (d, J = 16.3 Hz, 1H), 6.68 (d, J = 8.4 Hz, 2H), 3.73 (s, 2H). **¹³C NMR** (101 MHz, CDCl₃) δ 146.1, 137.9, 128.7, 128.6, 128.0, 127.7, 126.9, 126.1, 125.1, 115.2.

4-(Methylthio)aniline, 20l[217]

99%, dark brown solid. **¹H NMR** (600 MHz, CDCl₃) δ 7.19 (s, 2H), 6.63 (s, 2H), 3.75 (s, 2H), 2.42 (s, 3H). **¹³C NMR** (151 MHz, CDCl₃) δ 145.3, 131.0, 125.8, 115.9, 18.8.

4-Methoxyaniline, 20m[216]

88%, grey solid. **¹H NMR** (600 MHz, CDCl₃) δ 6.77 – 6.73 (m, 2H), 6.67 – 6.63 (m, 2H), 3.75 (s, 3H), 3.32 (s, 2H). **¹³C NMR** (151 MHz, CDCl₃) δ 152.8, 139.9, 116.4, 114.8, 55.7.

4-(Benzyloxy)aniline, 20n[216]

96%, brown solid. **¹H NMR** (400 MHz, CDCl₃) δ 7.47 – 7.27 (m, 5H), 6.82 (d, J = 8.8 Hz, 2H), 6.64 (d, J = 8.8 Hz, 2H), 4.99 (s, 2H), 3.43 (s, 2H). **¹³C NMR** (101 MHz, CDCl₃) δ 152.0, 140.2, 137.5, 128.5, 127.8, 127.5, 116.4, 116.1, 70.8.

Benzo[d][1,3]dioxol-5-amine, 20o[217]

96%, dark brown solid. **¹H NMR** (600 MHz, CDCl₃) δ 6.62 (d, J = 8.2 Hz, 1H), 6.29 (d, J = 2.3 Hz, 1H), 6.13 (dd, J = 8.2, 2.3 Hz, 1H), 5.86 (s, 2H), 3.47 (s, 2H). **¹³C NMR** (151 MHz, CDCl₃) δ 148.2, 141.4, 140.3, 108.6, 106.9, 100.7, 98.1.

Methyl 4-aminobenzoate, 20p[219]

79%, beige solid. **¹H NMR** (400 MHz, CDCl₃) δ 7.84 (d, J = 8.6 Hz, 2H), 6.63 (d, J = 8.6 Hz, 2H), 4.07 (s, 2H), 3.85 (s, 3H). **¹³C NMR** (101 MHz, CDCl₃) δ 167.2, 150.9, 131.6, 119.7, 113.8, 51.6.

Ethyl 2-(4-aminophenyl)acetate, 20q[222]

83%, brown oil. **¹H NMR** (600 MHz, CDCl₃) δ 6.99 (d, J = 8.3 Hz, 2H), 6.56 (d, J = 8.3 Hz, 2H), 4.05 (q, J = 7.1 Hz, 2H), 3.55 (s, 2H), 3.41 (s, 2H), 1.16 (t, J = 7.1 Hz, 3H). **¹³C NMR** (151 MHz, CDCl₃) δ 172.2, 145.4, 130.1, 124.0, 115.3, 60.7, 40.6, 14.2.

4-Amino-*N*-benzylbenzamide, 20r[223]

85%, beige solid. **¹H NMR** (600 MHz, CDCl₃) δ 7.62 (d, J = 8.1 Hz, 2H), 7.34 (d, J = 4.3 Hz, 4H), 7.30-7.27 (m, 1H), 6.65 (d, J = 8.1 Hz, 2H), 6.29 (s, 1H), 4.62 (d, J = 5.5 Hz, 2H), 3.98 (s, 2H). **¹³C NMR** (151 MHz, CDCl₃) δ 149.6, 138.6, 128.7, 127.9, 127.5, 123.9, 114.2, 44.0.

Benzene-1,4-diamine, 20s[216]

93%, dark brown solid. **¹H NMR** (400 MHz, CDCl₃) δ 6.56 (s, 4H), 3.28 (s, 4H). **¹³C NMR** (101 MHz, CDCl₃) δ 138.6, 116.7.

Benzene-1,2-diamine, 20t[224]

85%, grey solid. **¹H NMR** (600 MHz, CDCl₃) δ 6.63 (s, 4H), 3.38 (s, 4H). **¹³C NMR** (151 MHz, CDCl₃) δ 134.9, 120.3, 116.9.

2-Amino-*N*-benzylbenzenesulfonamide, 20u[225]

78%, brown oil. **¹H NMR** (600 MHz, CDCl₃) δ 7.75 (dd, J = 8.0, 1.4 Hz, 1H), 7.37 – 7.33 (m, 1H), 7.31 – 7.24 (m, 3H), 7.23 – 7.18 (m, 2H), 6.85 – 6.81 (m, 1H), 6.78 (d, J = 8.1 Hz, 1H), 5.10 (t, J = 5.7 Hz, 1H), 4.87 (s, 2H), 4.06 (d, J = 6.2 Hz, 2H). **¹³C NMR** (151 MHz, CDCl₃) δ 145.1, 136.3, 134.3, 129.8, 128.7, 127.9(89), 127.9(88), 121.5, 117.9, 117.8, 47.3.

Naphthalen-1-amine, 20v[216]

75%, dark brown solid. **¹H NMR** (400 MHz, CDCl₃) δ 7.76 – 7.68 (m, 2H), 7.42 – 7.33 (m, 2H), 7.26 – 7.14 (m, 2H), 6.69 (dd, J = 6.8, 1.5 Hz, 1H), 4.02 (s, 2H). **¹³C NMR** (101 MHz, CDCl₃) δ 142.1, 134.4, 128.6, 126.4, 125.9, 124.9, 123.7, 120.8, 119.0, 109.7.

4.6 Intermolecular alkylation of amines with alcohols to form heterocycles

4.6.1 General information

All reactions were carried out under an argon atmosphere using oven-dried glassware. Acetonitrile, THF and toluene were obtained from MBRAUN Solvent Purification System. The dry and degassed dioxane was distilled from sodium benzophenone under nitrogen. ¹H, ¹³C and ¹⁹F spectra were recorded in CDCl₃ using Varian VNMR 600 MHz, Ascend Bruker 600 MHz, Ascend Bruker 400 MHz and Inova 400 MHz spectrometer. The signals were referenced to residual chloroform (7.26 ppm, ^{1}H, 77.00 ppm, ^{13}C). Chemical shifts are reported in ppm, multiplicities are indicated by s (singlet), d (doublet), t (triplet), q (quartet), p (pentet or quintet), dd (doublet of doublets), td (triplet of doublets), br (broad signal) and m (multiplet). Analytical thin-layer chromatography (TLC) was performed using silica gel 60 pre-coated aluminium plates (Macherey-Nagel 0.20 mm thickness) with a fluorescent indicator UV254. Visualization was performed with standard phosphomolybdic acid stain (10g in 100 mL EtOH) or UV light. Analysis by gas chromatography (GC) was done using a CP-Sil-8-CB column (30 m, d = 0.25 mm) with FID detector and H$_2$ as carrier gas.

4.6.2 Experimental procedure and characterizations of the products

All chemicals, if not otherwise noted, were used as purchased without further purification.

Synthesis of starting materials

General procedures for a synthesis of compounds **24b-24f**

A solution of nitroarene (15 mmol, 1.0 equivalent), paraformaldehyde (15 mmol, 1.0 equiv.) and benzyltrimethylammonium hydroxide (Triton B, 40% solution of MeOH, 0.19 mL) in 2 mL DMSO was stirred at 90 °C for 2 h in the oil bath. Then, 15 mL of water was added and the aqueous phase was extracted with diethyl ether (3 ×25 mL).The combined organic extracts were dried over MgSO$_4$), filtered and concentrated under reduced pressure, giving a residue, which was purified by flash column chromatography (pentane:EtOAc=4:1) to afford pure products.

The hydrogenation of nitro group was performed using a procedure described in chapter 4.5.2. In an argon filled glovebox a 15 mL glass vial was charged with the corresponding intermediate (1.5 mmol), **Mn-6** (5 mol%), K$_2$CO$_3$ (12.5 mol%) and 3 mL of degassed toluene. The vial was sealed with a cap with a septum and was transferred into a stainless steel autoclave and a hole was made with a needle to allow an access of the gases. The autoclave was carefully flushed three times with nitrogen and then hydrogen gas. After adjusting the final hydrogen pressure to 80 bar, the autoclave was heated to 130 °C for 24 h with stirring. After cooling down the autoclave to room temperature the residual H$_2$ was carefully released and the mixture was analysed by TLC. Next, the reaction mixtures were purified by column chromatography on silica gel to obtain the pure amino alcohols **24**. The experimental data is in accordance with the reported in the literature **24b**,[226] **24c**,[227] **24d**,[227] **24e**,[227] **24f**.[227]

General procedures for a synthesis of compounds **26a-26g**

To the round bottom flask charged with PdCl$_2$(PPh$_3$)$_2$ (5 mol%) and CuI (1 mol%) anhydrous Et$_3$N was added (0.25M), followed by the addition of corresponding iodoarene (1 equiv.). Then, propargyl alcohol or but-3-yn-1-ol was added (1.5 equiv.) and obtained reaction mixture was stirred at room temperature until TLC shown full conversion of starting aniline. After filtration through the short pad of celite obtained filtrate was concentrated and obtained residue was purified by flash column chromatography (pentane:EtOAc=1:1). Next, the intermediate was placed in 15 mL glass vial, dissolved in ethanol, 5 mol% of Pd/C (10 wt%) was added and the reaction vessel was sealed and introduced to the autoclave. The autoclave was carefully flushed three times with nitrogen and then hydrogen gas. After adjusting the final hydrogen pressure to 20 bar, the autoclave was heated to 50 °C for 16 h with stirring. After cooling down the autoclave to room temperature the residual H$_2$ was carefully released and flushed few times with nitrogen prior to opening. Next, the reaction mixtures were purified by column chromatography on silica gel (pentane:EtOAc=1:1) to obtain the pure amino alcohol **26**. The experimental data is in accordance with the reported in the literature **26a**,[228] **26b**,[229] **26c**,[230] **26d**,[231] **26e**,[229] **26f**.[229]

General procedure for indoles synthesis

In an argon filled glovebox a glass pressure tube (10 mL) equipped with a magnetic stir bar was charged with Mn catalyst **Mn-6** (1.6 mg, 1 mol%) and KO^{t}Bu (0.7 mg, 2.5 mol%). degassed toluene (0.5 mL), and corresponding amino alcohol **24** (0.25 mmol) were added and the tube was closed with a screw cap. The resulting mixture was stirred at 120 °C for 24 h under argon atmosphere. After cooling down to room temperature the crude reaction mixture was purified by flash column chromatography on silica gel eluting with pentane/ethyl acetate (4:1) to give pure compounds **25**.

General procedure for a synthesis of 1,2,3,4-tetrahydroquinolines and 2,3,4,5-tetrahydro-1H-benzo[b]azepine

In an argon filled glovebox a glass pressure tube (10 mL) equipped with a magnetic stir bar was charged with Mn catalyst **Mn-6** (3.16 mg, 2 mol%) and KO^{t}Bu (28.1 mg, 1 equiv.) degassed toluene (0.5 mL), and corresponding amino alcohol **26** (0.25 mmol) were added and the tube was closed with a screw cap. The resulting mixture was stirred at 135 °C for 24 h under argon atmosphere. After cooling down to room temperature the crude reaction mixture was purified by flash column chromatography on silica gel eluting with pentane/ethyl acetate (4:1) to give pure compounds **27**.

Spectral data

Indole, 25a[208]

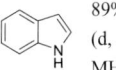
89%, grey solid. **^{1}H NMR** (600 MHz, CDCl$_3$) δ 8.06 (br, 1H), 7.58 (d, J = 7.9 Hz, 1H), 7.31 (d, J = 8.0 Hz, 1H), 7.15-7.08 (m, 2H), 7.04 (t, J = 7.4 Hz, 1H), 6.48 (s, 1H). **^{13}C NMR** (151 MHz, CDCl$_3$) δ 135.8, 127.9, 124.1, 122.0, 120.8, 119.8, 111.0, 102.6.

6-Bromo-1H-indole, 25b[232]

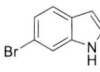
98%, grey solid. **^{1}H NMR** (400 MHz, CDCl$_3$) δ 8.16 (br, 1H), 7.58 (s, 1H), 7.53 (d, J = 8.4 Hz, 1H), 7.27 – 7.18 (m, 2H), 6.56 (s, 1H). **^{13}C NMR** (101 MHz, CDCl$_3$) δ 136.6, 126.8, 124.8, 123.2, 121.9, 115.5, 114.0, 102.9.

4-Chloro-1H-indole, 25c[231]

95%, light brown oil. **^{1}H NMR** (400 MHz, CDCl$_3$) δ 8.29 (br, 1H), 7.39 – 7.22 (m, 2H), 7.15 (s, 2H), 6.70 (s, 1H). **^{13}C NMR** (101 MHz, CDCl$_3$) δ 136.5, 126.8, 126.2, 124.7, 122.6, 119.6, 109.7, 101.4.

6-Chloro-1H-indole, 25d[233]

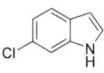
98%, grey solid. **^{1}H NMR** (400 MHz, CDCl$_3$) δ 8.16 (br, 1H), 7.57 (d, J = 8.4 Hz, 1H), 7.42 (s, 1H), 7.23 (s, 1H), 7.12 (d, J = 8.4 Hz, 1H), 6.56 (s, 1H). **^{13}C NMR** (101 MHz, CDCl$_3$) δ 136.2, 127.9, 126.5, 124.8, 121.6, 120.6, 111.0, 102.8.

5-Methoxy-1H-indole, 25e[231]

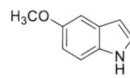
95%, light brown solid. **^{1}H NMR** (400 MHz, CDCl$_3$) δ 8.09 (br, 1H), 7.31 (d, J = 8.6 Hz, 1H), 7.21 (s, 1H), 7.15 (s, 1H), 6.91 (d, J = 8.3 Hz, 1H), 6.52 (s, 1H), 3.89 (s, 3H). **^{13}C NMR** (101 MHz, CDCl$_3$) δ 154.2, 131.0, 128.3, 124.9, 112.4, 111.7, 102.4(43), 102.4(38), 55.9.

3-methyl-1*H*-indole, 25f [208]

85%, white solid. **^{1}H NMR** (600 MHz, CDCl$_3$) δ 7.89 (s, 1H), 7.62 (d, J = 7.9 Hz, 1H), 7.38 (d, J = 8.1 Hz, 1H), 7.22 (t, J = 7.5 Hz, 1H), 7.16 (t, J = 7.4 Hz, 1H), 7.00 (s, 1H), 2.38 (s, 3H). **^{13}C NMR** (151 MHz, CDCl$_3$) δ 136.3, 128.3, 121.9, 121.5, 119.1, 118.8, 111.8, 110.9, 9.7.

1,2,3,4-Tetrahydroquinoline, 27a[234]

94%, dark red oil. **¹H NMR** (600 MHz, CDCl₃) δ 7.04 – 6.97 (m, 2H), 6.65 (t, J = 7.3 Hz, 1H), 6.51 (d, J = 7.9 Hz, 1H), 3.84 (br, 1H), 3.34 (t, J = 5.4 Hz, 2H), 2.81 (t, J = 6.3 Hz, 2H), 1.99 (p, J = 6.1 Hz, 2H). **¹³C NMR** (151 MHz, CDCl₃) δ 144.8, 129.5, 126.7, 121.5, 117.0, 114.2, 42.0, 27.0, 22.2.

6-Methyl-1,2,3,4-tetrahydroquinoline, 27b[235]

97%, light brown oil. **¹H NMR** (600 MHz, CDCl₃) δ 6.83-6.82 (m, 2H), 6.45 (d, J = 8.6 Hz, 1H), 3.57 (br, 1H), 3.31 (t, J = 5.4 Hz, 2H), 2.77 (t, J = 6.4 Hz, 2H), 2.25 (s, 3H), 1.97 (p, J = 6.2 Hz, 2H). **¹³C NMR** (151 MHz, CDCl₃) δ 142.4, 130.1, 127.3, 126.3, 121.6, 114.5, 42.2, 26.9, 22.5, 20.4.

7-Methoxy-1,2,3,4-tetrahydroquinoline, 27c[236]

98%, light brown solid. **¹H NMR** (600 MHz, CDCl₃) δ 6.88 (d, J = 8.2 Hz, 1H), 6.24 (dd, J = 8.2, 2.3 Hz, 1H), 6.07 (d, J = 2.3 Hz, 1H), 3.86 (br, 1H), 3.76 (s, 3H), 3.31 (t, J = 5.4 Hz, 2H), 2.73 (t, J = 6.4 Hz, 2H), 1.95 (p, J = 6.2 Hz, 2H). **¹³C NMR** (151 MHz, CDCl₃) δ 158.8, 145.6, 130.1, 114.1, 102.8, 99.5, 55.2, 42.0, 26.3, 22.4.

7-Chloro-1,2,3,4-tetrahydroquinoline, 27d[234]

98%, light brown oil. **¹H NMR** (600 MHz, CDCl₃) δ 6.86 (d, J = 8.0 Hz, 1H), 6.57 (dd, J = 8.0, 1.7 Hz, 1H), 6.47 (d, J = 1.6 Hz, 1H), 3.88 (br, 1H), 3.31 (t, J = 5.4 Hz, 2H), 2.73 (t, J = 6.4 Hz, 2H), 1.94 (p, J = 6.2 Hz, 2H). **¹³C NMR** (151 MHz, CDCl₃) δ 145.5, 131.9, 130.4, 119.7, 116.7, 113.5, 41.7, 26.5, 21.8.

6-Chloro-1,2,3,4-tetrahydroquinoline, 27e[234]

95%, brown oil. **¹H NMR** (600 MHz, CDCl₃) δ 6.91 (d, J = 9.0 Hz, 2H), 6.38 (d, J = 7.4 Hz, 1H), 3.83 (br, 1H), 3.28 (s, 2H), 2.73 (s, 2H), 1.91 (s, 2H). **¹³C NMR** (151 MHz, CDCl₃) δ 143.3, 129.0, 126.5, 122.9, 121.1, 115.1, 41.9, 26.9, 21.7.

2,3,4,5-tetrahydro-1H-benzo[b]azepine, 27f[237]

84%, light brown oil. **¹H NMR** (600 MHz, CDCl₃) δ 7.14 (d, J = 7.2 Hz, 1H), 7.07 (t, J = 7.0 Hz, 1H), 6.86 (t, J = 7.2 Hz, 1H), 6.76 (d, J = 7.3 Hz, 1H), 3.80 (br, 1H), 3.09 (s, 2H), 2.81 (s, 2H), 1.84 (s, 2H), 1.68 (s, 2H). **¹³C NMR** (151 MHz, CDCl₃) δ 150.5, 133.8, 130.8, 126.6, 120.9, 119.4, 49.0, 36.1, 32.0, 26.9.

4.7 Manganese-catalysed alkylation of nitroarenes with alcohols

4.7.1 General information

All reactions were carried out under an argon atmosphere using oven-dried glassware. The dry and degassed methanol was distilled from calcium hydride under nitrogen. All other chemicals were used as purchased without further purification. ^{1}H and ^{13}C spectra were recorded in CDCl$_3$ using Varian VNMR 600 MHz and Ascend Bruker 600 MHz. The signals were referenced to residual chloroform (7.26 ppm, ^{1}H, 77.00 ppm, ^{13}C). Chemical shifts are reported in ppm, multiplicities are indicated by s (singlet), d (doublet), t (triplet), p (pentet or quintet), br (broad signal) and m (multiplcet). Analytical thin-layer chromatography (TLC) was performed using silica gel 60 pre-coated aluminium plates (Macherey-Nagel 0.20 mm thickness) with a fluorescent indicator UV254. Visualization was performed with standard phosphomolybdic acid stain (10g in 100 mL EtOH) or UV light. Analysis by gas chromatography (GC) was done using a CP-Sil-8-CB column (30 m, d = 0.25 mm) with FID detector and H$_2$ as carrier gas.

4.7.2 Experimental procedure and characterizations of the products

General procedure for an alkylation of nitroarenes with alcohols

In an argon filled glovebox a glass pressure tube (10 mL) equipped with a magnetic stir bar was charged with Mn catalyst **Mn-6** (9.5 mg, 3 mol%) and Cs$_2$CO$_3$ (326 mg, 2 equiv.). selected alcohol **32** (2 mL), and corresponding nitroarene **28** (0.5 mmol) were added and the tube was closed with a screw cap. The resulting mixture was stirred at 140 °C for 24 h under argon atmosphere. After cooling down to room temperature the crude reaction mixture was purified by flash column chromatography on silica gel eluting with pentane to give pure compounds **30** and **31**.

Spectral data

N-methylaniline, 30a[238]

82%, light brown oil. ^{1}**H NMR** (600 MHz, CDCl$_3$) δ 7.25 (t, *J* = 7.9 Hz, 2H), 6.77 (t, *J* = 7.3 Hz, 1H), 6.67 (d, *J* = 7.8 Hz, 2H), 3.70 (br, 1H), 2.88 (s, 3H). ^{13}**C NMR** (151 MHz, CDCl$_3$) δ 149.4, 129.2, 117.3, 112.5, 30.8.

N,4-dimethylaniline, 30b[72]

94%, light brown oil. ^{1}**H NMR** (600 MHz, CDCl$_3$) δ 6.93 (d, *J* = 8.2 Hz, 2H), 6.47 (d, *J* = 8.4 Hz, 2H), 3.42 (br, 1H), 2.73 (s, 3H), 2.17 (s, 3H). ^{13}**C NMR** (151 MHz, CDCl$_3$) δ 147.2, 129.7, 126.5, 112.6, 31.1, 20.4.

4-(*tert*-Butyl)-*N*-methylaniline, 30c[72]

97%, brown oil. ^{1}**H NMR** (600 MHz, CDCl$_3$) δ 7.33 – 7.27 (m, 2H), 6.67 – 6.62 (m, 2H), 3.62 (br, 1H), 2.88 (s, 3H), 1.36 (s, 9H). ^{13}**C NMR** (151 MHz, CDCl$_3$) δ 147.1, 140.1, 126.0, 112.3, 33.9, 31.6, 31.0.

N-butyl-4-methylaniline, 31a[239]

69%, brown oil. **¹H NMR** (600 MHz, CDCl₃) δ 7.02 (d, *J* = 8.1 Hz, 2H), 6.57 (d, *J* = 8.4 Hz, 2H), 3.52 (br, 1H), 3.12 (t, *J* = 7.1 Hz, 2H), 2.27 (s, 3H), 1.65-1.60 (m, 2H), 1.49-1.43 (m, 2H), 0.99 (t, *J* = 7.4 Hz, 3H). **¹³C NMR** (151 MHz, CDCl₃) δ 146.3, 129.7, 126.3, 112.9, 44.1, 31.8, 20.4, 20.3, 13.9.

2-(p-Tolylamino)ethan-1-ol, 31b[240]

81%, light brown oil. **¹H NMR** (600 MHz, CDCl₃) δ 6.92 (d, *J* = 8.0 Hz, 2H), 6.50 (d, *J* = 8.3 Hz, 2H), 3.72 (s, 2H), 3.19 (s, 2H), 2.17 (s, 3H). **¹³C NMR** (151 MHz, CDCl₃) δ 145.9, 129.8, 127.3, 113.6, 61.3, 46.6, 20.4.

N-hexyl-4-methylaniline, 31c[240]

60%, light brown oil. **¹H NMR** (600 MHz, CDCl₃) δ 6.90 (d, *J* = 8.1 Hz, 2H), 6.45 (d, *J* = 8.4 Hz, 2H), 3.36 (s, 1H), 2.99 (t, *J* = 7.2 Hz, 2H), 2.15 (s, 3H), 1.52 (p, *J* = 7.3 Hz, 2H), 1.34-1.28 (m, 2H), 1.27-1.20 (m, 4H), 0.82 (t, *J* = 7.0 Hz, 3H). **¹³C NMR** (151 MHz, CDCl₃) δ 146.4, 129.7, 126.3, 112.9, 44.4, 31.7, 29.7, 26.9, 22.7, 20.4, 14.1.

Chapter 5: List of abbreviations

AD	Asseptorless dehydrogenation
ADC	Asseptorless dehydrogenative coupling
Alk	Alkyl
Ar	Aryl
ATR	Attenuated total reflectance
Bn	Benzyl
Bu	Butyl
COCs	Cyclic organic carbonates
DCM	Dichloromethane
DFT	Density functional theory
DG	Directing group
DMF	Dimethylformamide
dr	Diastereomeric ratio
EG	Ethylene glycol
EI	Electron ionization
equiv.	Equivalent
ESI	Electrospray ionization
Et	Ethyl
EtOH	Ethanol
FID	Flame ionization detector
g	Gram
GC	Gas chromatography
h	Hours
HA	Hydrogen autotransfer
Het	Heterocycle
HRMS	High resolution mass spectrometry
Hz	Hertz
iPr	Isopropyl
IR	Infrared spectroscopy
J	Coupling constant
kg	Kilogram
L	Ligand
LOHC	Liquid organic hydrogen carrier
M	Metal
m/z	Value of mass divided by charge
M+	Molecular ion
Me	Methyl
MeOH	Methanol
mL	Mililiter
MS	Mass spectrometry
NMP	*N*-Methyl-2-pyrrolidone
NMR	Nuclear magnetic resonance spectroscopy
PES	Potential energy surface
Ph	Phenyl
PMP	*para*-Methoxyphenyl
ppm	Part per million
rac	Racemic
rt	Room temperature

TAA	*t*-Amyl alcohol
TBS	*t*-Butyldimethylsilyl
^{t}Bu	*t*-Butyl
TH	Transfer hydrogenation
THF	Tetrahydrofuran
TIPS	Triisopropylsilyl
TLC	Thin-layer chromatography
TS	Transition state
UV	Ultraviolet radiation

Chapter 6: References

[1] E. C. Constable, C. E. Housecroft, *Chemical Society reviews* **2013**, *42*, 1429.
[2] R. Franke, D. Selent, A. Börner, *Chemical reviews* **2012**, *112*, 5675.
[3] J. Smidt, W. Hafner, R. Jira, J. Sedlmeier, R. Sieber, R. Rüttinger, H. Kojer, *Angew. Chem.* **1959**, *71*, 176.
[4] Study on the review of the list of critical raw materials. Final report, Publications Office of the European Union, Luxembourg, **2017**.
[5] European Medicines Agency, *GUIDELINE FOR ELEMENTAL IMPURITIES*, **2019**.
[6] M. D. Wodrich, X. Hu, *Nat Rev Chem* **2018**, *2*.
[7] a) S. Hashiguchi, A. Fujii, J. Takehara, T. Ikariya, R. Noyori, *J. Am. Chem. Soc.* **1995**, *117*, 7562; b) T. Ohkuma, H. Ooka, S. Hashiguchi, T. Ikariya, R. Noyori, *J. Am. Chem. Soc.* **1995**, *117*, 2675; c) T. Ohkuma, H. Ooka, T. Ikariya, R. Noyori, *J. Am. Chem. Soc.* **1995**, *117*, 10417.
[8] a) P. E. Sues, K. Z. Demmans, R. H. Morris, *Dalton transactions (Cambridge, England : 2003)* **2014**, *43*, 7650; b) W. Zuo, A. J. Lough, Y. F. Li, R. H. Morris, *Science (New York, N.Y.)* **2013**, *342*, 1080.
[9] E. Peris, R. H. Crabtree, *Chemical Society reviews* **2018**, *47*, 1959.
[10] M. D. Fryzuk, P. A. MacNeil, *Organometallics* **1983**, *2*, 682.
[11] a) B. Askevold, H. W. Roesky, S. Schneider, *ChemCatChem* **2012**, *4*, 307; b) S. Schneider, J. Meiners, B. Askevold, *Eur. J. Inorg. Chem.* **2012**, *2012*, 412.
[12] P. O. Lagaditis, P. E. Sues, J. F. Sonnenberg, K. Y. Wan, A. J. Lough, R. H. Morris, *J. Am. Chem. Soc.* **2014**, *136*, 1367.
[13] S. Chakraborty, P. O. Lagaditis, M. Förster, E. A. Bielinski, N. Hazari, M. C. Holthausen, W. D. Jones, S. Schneider, *ACS Catal.* **2014**, *4*, 3994.
[14] X. Chen, W. Jia, R. Guo, T. W. Graham, M. A. Gullons, K. Abdur-Rashid, *Dalton transactions (Cambridge, England : 2003)* **2009**, 1407.
[15] a) C. Bornschein, S. Werkmeister, B. Wendt, H. Jiao, E. Alberico, W. Baumann, H. Junge, K. Junge, M. Beller, *Nature communications* **2014**, *5*, 4111; b) J. Neumann, C. Bornschein, H. Jiao, K. Junge, M. Beller, *Eur. J. Org. Chem.* **2015**, *2015*, 5944.
[16] a) T. J. Schmeier, G. E. Dobereiner, R. H. Crabtree, N. Hazari, *J. Am. Chem. Soc.* **2011**, *133*, 9274; b) Y. Zhang, A. D. MacIntosh, J. L. Wong, E. A. Bielinski, P. G. Williard, B. Q. Mercado, N. Hazari, W. H. Bernskoetter, *Chemical science* **2015**, *6*, 4291.
[17] a) S. Chakraborty, H. Dai, P. Bhattacharya, N. T. Fairweather, M. S. Gibson, J. A. Krause, H. Guan, *J. Am. Chem. Soc.* **2014**, *136*, 7869; b) S. Werkmeister, K. Junge, B. Wendt, E. Alberico, H. Jiao, W.

Baumann, H. Junge, F. Gallou, M. Beller, *Angewandte Chemie (International ed. in English)* **2014**, *53*, 8722.

[18] D. Spasyuk, S. Smith, D. G. Gusev, *Angewandte Chemie (International ed. in English)* **2012**, *51*, 2772.

[19] Z. Han, L. Rong, J. Wu, L. Zhang, Z. Wang, K. Ding, *Angewandte Chemie (International ed. in English)* **2012**, *51*, 13041.

[20] S. Chakraborty, W. W. Brennessel, W. D. Jones, *J. Am. Chem. Soc.* **2014**, *136*, 8564.

[21] a) M. Bertoli, A. Choualeb, A. J. Lough, B. Moore, D. Spasyuk, D. G. Gusev, *Organometallics* **2011**, *30*, 3479; b) M. Nielsen, A. Kammer, D. Cozzula, H. Junge, S. Gladiali, M. Beller, *Angewandte Chemie (International ed. in English)* **2011**, *50*, 9593.

[22] E. A. Bielinski, P. O. Lagaditis, Y. Zhang, B. Q. Mercado, C. Würtele, W. H. Bernskoetter, N. Hazari, S. Schneider, *J. Am. Chem. Soc.* **2014**, *136*, 10234.

[23] a) M. Käss, A. Friedrich, M. Drees, S. Schneider, *Angewandte Chemie (International ed. in English)* **2009**, *48*, 905; b) A. N. Marziale, A. Friedrich, I. Klopsch, M. Drees, V. R. Celinski, J. Schmedt auf der Günne, S. Schneider, *J. Am. Chem. Soc.* **2013**, *135*, 13342.

[24] C. Gunanathan, D. Milstein, *Accounts of chemical research* **2011**, *44*, 588.

[25] D. E. Prokopchuk, B. T. H. Tsui, A. J. Lough, R. H. Morris, *Chemistry (Weinheim an der Bergstrasse, Germany)* **2014**, *20*, 16960.

[26] a) C. A. Huff, J. W. Kampf, M. S. Sanford, *Organometallics* **2012**, *31*, 4643; b) C. A. Huff, J. W. Kampf, M. S. Sanford, *Chemical communications (Cambridge, England)* **2013**, *49*, 7147.

[27] a) G. A. Filonenko, E. Cosimi, L. Lefort, M. P. Conley, C. Copéret, M. Lutz, E. J. M. Hensen, E. A. Pidko, *ACS Catal.* **2014**, *4*, 2667; b) G. A. Filonenko, D. Smykowski, B. M. Szyja, G. Li, J. Szczygieł, E. J. M. Hensen, E. A. Pidko, *ACS Catal.* **2015**, *5*, 1145.

[28] a) F. Franco, C. Cometto, F. Ferrero Vallana, F. Sordello, E. Priola, C. Minero, C. Nervi, R. Gobetto, *Chemical communications (Cambridge, England)* **2014**, *50*, 14670; b) M. D. Sampson, A. D. Nguyen, K. A. Grice, C. E. Moore, A. L. Rheingold, C. P. Kubiak, *J. Am. Chem. Soc.* **2014**, *136*, 5460; c) J. M. Smieja, M. D. Sampson, K. A. Grice, E. E. Benson, J. D. Froehlich, C. P. Kubiak, *Inorganic chemistry* **2013**, *52*, 2484; d) H. Takeda, K. Koizumi, K. Okamoto, O. Ishitani, *Chemical communications (Cambridge, England)* **2014**, *50*, 1491.

[29] D. A. Valyaev, G. Lavigne, N. Lugan, *Coordination Chemistry Reviews* **2016**, *308*, 191.

[30] K. Srinivasan, P. Michaud, J. K. Kochi, *J. Am. Chem. Soc.* **1986**, *108*, 2309.

[31] W. Zhang, J. L. Loebach, S. R. Wilson, E. N. Jacobsen, *J. Am. Chem. Soc.* **1990**, *112*, 2801.

[32] R. Irie, K. Noda, Y. Ito, T. Katsuki, *Tetrahedron Letters* **1991**, *32*, 1055.

[33] M. I. Bruce, M. Z. Iqbal, F. G. A. Stone, *J. Chem. Soc., A* **1970**, 3204.

[34] a) R. He, Z.-T. Huang, Q.-Y. Zheng, C. Wang, *Angewandte Chemie (International ed. in English)* **2014**, *53*, 4950; b) Y. Kuninobu, Y. Nishina, T. Takeuchi, K. Takai, *Angewandte Chemie (International ed. in English)* **2007**, *46*, 6518; c) T. Sato, T. Yoshida, H. H. Al Mamari, L. Ilies, E. Nakamura, *Organic letters* **2017**, *19*, 5458; d) S. Sueki, Z. Wang, Y. Kuninobu, *Organic letters* **2016**, *18*, 304; e) B. Zhou, H. Chen, C. Wang, *J. Am. Chem. Soc.* **2013**, *135*, 1264; f) B. Zhou, Y. Hu, C. Wang, *Angewandte Chemie (International ed. in English)* **2015**, *54*, 13659; g) B. Zhou, P. Ma, H. Chen, C. Wang, *Chemical communications (Cambridge, England)* **2014**, *50*, 14558.

[35] C. Wang, B. Maity, L. Cavallo, M. Rueping, *Organic letters* **2018**, *20*, 3105.

[36] a) M. L. Clarke, *Catal. Sci. Technol.* **2012**, *2*, 2418; b) P. A. Dub, T. Ikariya, *ACS Catal.* **2012**, *2*, 1718; c) S. Werkmeister, K. Junge, M. Beller, *Org. Process Res. Dev.* **2014**, *18*, 289.

[37] R. M. Bullock, *Catalysis without precious metals*, Wiley-VCH, Weinheim, **2010**.

[38] S. Elangovan, C. Topf, S. Fischer, H. Jiao, A. Spannenberg, W. Baumann, R. Ludwig, K. Junge, M. Beller, *J. Am. Chem. Soc.* **2016**, *138*, 8809.

[39] F. Kallmeier, T. Irrgang, T. Dietel, R. Kempe, *Angewandte Chemie (International ed. in English)* **2016**, *55*, 11806.

[40] a) A. Bruneau-Voisine, D. Wang, T. Roisnel, C. Darcel, J.-B. Sortais, *Catalysis Communications* **2017**, *92*, 1; b) D. Wei, A. Bruneau-Voisine, T. Chauvin, V. Dorcet, T. Roisnel, D. A. Valyaev, N. Lugan, J.-B. Sortais, *Adv. Synth. Catal.* **2018**, *360*, 676.

[41] S. Weber, B. Stöger, K. Kirchner, *Organic letters* **2018**, *20*, 7212.

[42] M. Glatz, B. Stöger, D. Himmelbauer, L. F. Veiros, K. Kirchner, *ACS Catal.* **2018**, *8*, 4009.

[43] a) M. Garbe, K. Junge, S. Walker, Z. Wei, H. Jiao, A. Spannenberg, S. Bachmann, M. Scalone, M. Beller, *Angewandte Chemie (International ed. in English)* **2017**, *56*, 11237; b) M. Garbe, Z. Wei, B. Tannert, A. Spannenberg, H. Jiao, S. Bachmann, M. Scalone, K. Junge, M. Beller, *Adv. Synth. Catal.* **2019**, *361*, 1913; c) F. Ling, H. Hou, J. Chen, S. Nian, X. Yi, Z. Wang, D. Song, W. Zhong, *Organic letters* **2019**, *21*, 3937; d) M. B. Widegren, G. J. Harkness, A. M. Z. Slawin, D. B. Cordes, M. L. Clarke, *Angewandte Chemie (International ed. in English)* **2017**, *56*, 5825; e) L. Zhang, Y. Tang, Z. Han, K. Ding, *Angewandte Chemie (International ed. in English)* **2019**, *58*, 4973.

[44] J. A. Garduño, J. J. García, *ACS Catal.* **2019**, *9*, 392.

[45] S. Elangovan, M. Garbe, H. Jiao, A. Spannenberg, K. Junge, M. Beller, *Angew. Chem. Int. Ed.* **2016**, *55*, 15364.

[46] N. A. Espinosa-Jalapa, A. Nerush, L. J. W. Shimon, G. Leitus, L. Avram, Y. Ben-David, D. Milstein, *Chemistry (Weinheim an der Bergstrasse, Germany)* **2017**, *23*, 5934.

[47] R. van Putten, E. A. Uslamin, M. Garbe, C. Liu, A. Gonzalez-de-Castro, M. Lutz, K. Junge, E. J. M. Hensen, M. Beller, L. Lefort et al., *Angewandte Chemie (International ed. in English)* **2017**, *56*, 7531.

[48] M. B. Widegren, M. L. Clarke, *Organic letters* **2018**, *20*, 2654.
[49] V. Papa, J. R. Cabrero-Antonino, E. Alberico, A. Spanneberg, K. Junge, H. Junge, M. Beller, *Chemical science* **2017**, *8*, 3576.
[50] Y.-Q. Zou, S. Chakraborty, A. Nerush, D. Oren, Y. Diskin-Posner, Y. Ben-David, D. Milstein, *ACS Catal.* **2018**, *8*, 8014.
[51] C. Gunanathan, D. Milstein, *Science (New York, N.Y.)* **2013**, *341*, 1229712.
[52] M. Andérez-Fernández, L. K. Vogt, S. Fischer, W. Zhou, H. Jiao, M. Garbe, S. Elangovan, K. Junge, H. Junge, R. Ludwig et al., *Angew. Chem.* **2017**, *129*, 574.
[53] A. M. Tondreau, J. M. Boncella, *Organometallics* **2016**, *35*, 2049.
[54] D. H. Nguyen, Y. Morin, L. Zhang, X. Trivelli, F. Capet, S. Paul, S. Desset, F. Dumeignil, R. M. Gauvin, *ChemCatChem* **2017**, *9*, 2652.
[55] S. Chakraborty, U. Gellrich, Y. Diskin-Posner, G. Leitus, L. Avram, D. Milstein, *Angew. Chem. Int. Ed.* **2017**, *56*, 4229.
[56] A. Kumar, N. A. Espinosa-Jalapa, G. Leitus, Y. Diskin-Posner, L. Avram, D. Milstein, *Angew. Chem. Int. Ed.* **2017**, *56*, 14992.
[57] D. H. Nguyen, X. Trivelli, F. Capet, J.-F. Paul, F. Dumeignil, R. M. Gauvin, *ACS Catal.* **2017**, *7*, 2022.
[58] N. A. Espinosa-Jalapa, A. Kumar, G. Leitus, Y. Diskin-Posner, D. Milstein, *J. Am. Chem. Soc.* **2017**, *139*, 11722.
[59] A. Mukherjee, A. Nerush, G. Leitus, L. J. W. Shimon, Y. Ben David, N. A. Espinosa Jalapa, D. Milstein, *J. Am. Chem. Soc.* **2016**, *138*, 4298.
[60] M. Mastalir, M. Glatz, N. Gorgas, B. Stöger, E. Pittenauer, G. Allmaier, L. F. Veiros, K. Kirchner, *Chemistry (Weinheim an der Bergstrasse, Germany)* **2016**, *22*, 12316.
[61] J. O. Bauer, S. Chakraborty, D. Milstein, *ACS Catal.* **2017**, *7*, 4462.
[62] S. Chakraborty, U. K. Das, Y. Ben-David, D. Milstein, *J. Am. Chem. Soc.* **2017**, *139*, 11710.
[63] M. Mastalir, M. Glatz, E. Pittenauer, G. Allmaier, K. Kirchner, *J. Am. Chem. Soc.* **2016**, *138*, 15543.
[64] N. Deibl, R. Kempe, *Angew. Chem. Int. Ed.* **2017**, *56*, 1663.
[65] F. Kallmeier, B. Dudziec, T. Irrgang, R. Kempe, *Angew. Chem. Int. Ed.* **2017**, *56*, 7261.
[66] G. Guillena, D. J Ramón, M. Yus, *Chemical reviews* **2010**, *110*, 1611.
[67] A. J. A. Watson, J. M. J. Williams, *Science (New York, N.Y.)* **2010**, *329*, 635.
[68] M. H. S. A. Hamid, P. A. Slatford, J. M. J. Williams, *Adv. Synth. Catal.* **2007**, *349*, 1555.
[69] M. Perez, S. Elangovan, A. Spannenberg, K. Junge, M. Beller, *ChemSusChem* **2017**, *10*, 83.
[70] A. Bruneau-Voisine, D. Wang, V. Dorcet, T. Roisnel, C. Darcel, J.-B. Sortais, *Organic letters* **2017**, *19*, 3656.

[71] A. Zirakzadeh, S. R. M. M. de Aguiar, B. Stöger, M. Widhalm, K. Kirchner, *ChemCatChem* **2017**, *9*, 1744.

[72] S. Elangovan, J. Neumann, J.-B. Sortais, K. Junge, C. Darcel, M. Beller, *Nature communications* **2016**, *7*, 12641.

[73] J. Neumann, S. Elangovan, A. Spannenberg, K. Junge, M. Beller, *Chemistry (Weinheim an der Bergstrasse, Germany)* **2017**, *23*, 5410.

[74] A. Bruneau-Voisine, D. Wang, V. Dorcet, T. Roisnel, C. Darcel, J.-B. Sortais, *Journal of Catalysis* **2017**, *347*, 57.

[75] M. Peña-López, P. Piehl, S. Elangovan, H. Neumann, M. Beller, *Angew. Chem. Int. Ed.* **2016**, *55*, 14967.

[76] S. Fu, Z. Shao, Y. Wang, Q. Liu, *J. Am. Chem. Soc.* **2017**, *139*, 11941.

[77] A. Goeppert, M. Czaun, J.-P. Jones, G. K. Surya Prakash, G. A. Olah, *Chemical Society reviews* **2014**, *43*, 7995.

[78] a) E. Balaraman, Y. Ben-David, D. Milstein, *Angewandte Chemie (International ed. in English)* **2011**, *50*, 11702; b) E. Balaraman, C. Gunanathan, J. Zhang, L. J. W. Shimon, D. Milstein, *Nature chemistry* **2011**, *3*, 609.

[79] J. Schneidewind, R. Adam, W. Baumann, R. Jackstell, M. Beller, *Angewandte Chemie (International ed. in English)* **2017**, *56*, 1890.

[80] S. Kar, A. Goeppert, J. Kothandaraman, G. K. S. Prakash, *ACS Catal.* **2017**, *7*, 6347.

[81] a) M. Cokoja, M. E. Wilhelm, M. H. Anthofer, W. A. Herrmann, F. E. Kühn, *ChemSusChem* **2015**, *8*, 2436; b) J. W. Comerford, I. D. V. Ingram, M. North, X. Wu, *Green Chem.* **2015**, *17*, 1966; c) Q. He, J. W. O'Brien, K. A. Kitselman, L. E. Tompkins, G. C. T. Curtis, F. M. Kerton, *Catal. Sci. Technol.* **2014**, *4*, 1513; d) C. Martín, G. Fiorani, A. W. Kleij, *ACS Catal.* **2015**, *5*, 1353; e) M. North, R. Pasquale, C. Young, *Green Chem.* **2010**, *12*, 1514.

[82] E. Alper, O. Yuksel Orhan, *Petroleum* **2017**, *3*, 109.

[83] G. A. Filonenko, R. van Putten, E. J. M. Hensen, E. A. Pidko, *Chemical Society reviews* **2018**, *47*, 1459.

[84] M. Garbe, K. Junge, M. Beller, *Eur. J. Org. Chem.* **2017**, 4344.

[85] N. Gorgas, K. Kirchner, *Accounts of chemical research* **2018**, *51*, 1558.

[86] F. Kallmeier, R. Kempe, *Angew. Chem. Int. Ed.* **2018**, *57*, 46.

[87] B. Maji, M. Barman, *Synthesis* **2017**, *49*, 3377.

[88] T. Zell, R. Langer, *ChemCatChem* **2018**, *10*, 1930.

[89] K. Nakano, S. Hashimoto, M. Nakamura, T. Kamada, K. Nozaki, *Angewandte Chemie (International ed. in English)* **2011**, *50*, 4868.

[90] *Optimised geometries were located using the wB97XD functional together with the SVP basis set for main-group atoms and the TZVP basis set for Mn. To obtain more accurate energy values, single-point refinement calculations were done using the M06 functional and TZVP basis set on all atoms. Solvent effects (1,4-dioxane) were included with the PCM model at both steps. All calculations were performed through the facilities provided by the Gaussian09 package. See the Supporting Information for full computational details.*

[91] a) P. A. Dub, N. J. Henson, R. L. Martin, J. C. Gordon, *J. Am. Chem. Soc.* **2014**, *136*, 3505; b) F. Hasanayn, A. Baroudi, A. A. Bengali, A. S. Goldman, *Organometallics* **2013**, *32*, 6969; c) F. Hasanayn, R. H. Morris, *Inorganic chemistry* **2012**, *51*, 10808.

[92] A. Kumar, T. Janes, N. A. Espinosa-Jalapa, D. Milstein, *Angewandte Chemie (International ed. in English)* **2018**, *57*, 12076.

[93] A. Kaithal, M. Hölscher, W. Leitner, *Angewandte Chemie (International ed. in English)* **2018**, *57*, 13449.

[94] M. Breuer, K. Ditrich, T. Habicher, B. Hauer, M. Kesseler, R. Stürmer, T. Zelinski, *Angew. Chem. Int. Ed.* **2004**, *43*, 788.

[95] T. C. Nugent, M. El-Shazly, *Adv. Synth. Catal.* **2010**, *352*, 753.

[96] J.-H. Xie, S.-F. Zhu, Q.-L. Zhou, *Chemical reviews* **2011**, *111*, 1713.

[97] a) N. Fleury-Brégeot, V. de la Fuente, S. Castillón, C. Claver, *ChemCatChem* **2010**, *2*, 1346; b) C. Wang, B. Villa-Marcos, J. Xiao, *Chemical communications (Cambridge, England)* **2011**, *47*, 9773.

[98] S. Bähn, S. Imm, L. Neubert, M. Zhang, H. Neumann, M. Beller, *ChemCatChem* **2011**, *3*, 1853.

[99] a) G. Chelucci, *Coordination Chemistry Reviews* **2017**, *331*, 1; b) A. Corma, J. Navas, M. J. Sabater, *Chemical reviews* **2018**, *118*, 1410; c) A. Nandakumar, S. P. Midya, V. G. Landge, E. Balaraman, *Angew. Chem. Int. Ed.* **2015**, *54*, 11022; d) Y. Obora, *ACS Catal.* **2014**, *4*, 3972; e) A. Quintard, J. Rodriguez, *Chemical communications (Cambridge, England)* **2016**, *52*, 10456.

[100] J. Leonard, A. J. Blacker, S. P. Marsden, M. F. Jones, K. R. Mulholland, R. Newton, *Org. Process Res. Dev.* **2015**, *19*, 1400.

[101] Q. Yang, Q. Wang, Z. Yu, *Chemical Society reviews* **2015**, *44*, 2305.

[102] a) C. S. Lim, T. T. Quach, Y. Zhao, *Angew. Chem. Int. Ed.* **2017**, *56*, 7176; b) Z.-Q. Rong, Y. Zhang, R. H. B. Chua, H.-J. Pan, Y. Zhao, *J. Am. Chem. Soc.* **2015**, *137*, 4944.

[103] Y. Zhang, C.-S. Lim, D. S. B. Sim, H.-J. Pan, Y. Zhao, *Angew. Chem. Int. Ed.* **2014**, *53*, 1399.

[104] a) L.-C. Yang, Y.-N. Wang, Y. Zhang, Y. Zhao, *ACS Catal.* **2017**, *7*, 93; b) A. Eka Putra, Y. Oe, T. Ohta, *Eur. J. Org. Chem.* **2013**, *2013*, 6146.

[105] M. Peña-López, H. Neumann, M. Beller, *Angew. Chem. Int. Ed.* **2016**, *55*, 7826.

[106] P. Yang, C. Zhang, Y. Ma, C. Zhang, A. Li, B. Tang, J. S. Zhou, *Angew. Chem. Int. Ed.* **2017**, *56*, 14702.

[107] N. J. Oldenhuis, V. M. Dong, Z. Guan, *J. Am. Chem. Soc.* **2014**, *136*, 12548.

[108] G. Liu, D. A. Cogan, J. A. Ellman, *J. Am. Chem. Soc.* **1997**, *119*, 9913.

[109] M. T. Robak, M. A. Herbage, J. A. Ellman, *Chemical reviews* **2010**, *110*, 3600.

[110] J. C. Borghs, L. M. Azofra, T. Biberger, O. Linnenberg, L. Cavallo, M. Rueping, O. El-Sepelgy, *ChemSusChem* **2019**, *12*, 3083.

[111] J. C. Borghs, Y. Lebedev, M. Rueping, O. El-Sepelgy, *Organic letters* **2019**, *21*, 70.

[112] G. Tsuji, T. Takeda, I. Furusawa, O. Horino, Y. Kubo, *Pesticide Biochemistry and Physiology* **1997**, *57*, 211.

[113] C. M. Spencer, S. Noble, *Drugs & aging* **1998**, *13*, 391.

[114] S. A. Lawrence, *Amines. Synthesis, properties and applications / Stephen A. Lawrence*, Cambridge University Press, Cambridge, **2004**.

[115] a) T. C. Nugent, *Chiral amine synthesis. Methods, developments and applications / edited by Thomas C. Nugent*, Wiley-VCH, Weinheim, **2010**; b) A. Ricci, *Modern amination methods*, Wiley-VCH, Weinheim, Cambridge, **2000**.

[116] D. Kucera, R. Scott, US20040204591A1.

[117] J. R. Khusnutdinova, D. Milstein, *Angewandte Chemie (International ed. in English)* **2015**, *54*, 12236.

[118] V. Lyaskovskyy, B. de Bruin, *ACS Catal.* **2012**, *2*, 270.

[119] O. R. Luca, R. H. Crabtree, *Chemical Society reviews* **2013**, *42*, 1440.

[120] a) J. S. M. Samec, J.-E. Bäckvall, P. G. Andersson, P. Brandt, *Chemical Society reviews* **2006**, *35*, 237; b) S. E. Clapham, A. Hadzovic, R. H. Morris, *Coordination Chemistry Reviews* **2004**, *248*, 2201.

[121] Y. Zhao, S. W. Foo, S. Saito, *Angew. Chem. Int. Ed.* **2011**, *50*, 3006.

[122] H.-J. Pan, Y. Zhang, C. Shan, Z. Yu, Y. Lan, Y. Zhao, *Angew. Chem. Int. Ed.* **2016**, *55*, 9615.

[123] P. N. Rylander, *Catalytic hydrogenation in organic syntheses*, Academic Press, New York, **1979**.

[124] J. G. de Vries, C. J. Elsevier, *The handbook of homogeneous hydrogenation*, Wiley-VCH, Weinheim, Great Britain, **2007**.

[125] J. M. J. Williams, *Preparation of alkenes. A practical approach / edited by Jonathan M. J. Williams*, Oxford University Press, Oxford, **1996**.

[126] a) A. Fürstner, *Angew. Chem.* **2000**, *39*, 3012; b) L. Horner, H. Hoffmann, H. G. Wippel, *Chem. Ber.* **1958**, *91*, 61; c) M. Julia, J.-M. Paris, *Tetrahedron Letters* **1973**, *14*, 4833; d) P. J. Kocienski, B. Lythgoe, S. Ruston, *J. Chem. Soc., Perkin Trans. 1* **1978**, 829; e) D. J. Peterson, *J. Org. Chem.* **1968**, *33*, 780; f) R. R. Schrock, *Chemical reviews* **2002**, *102*, 145; g) W.C. Still, C. Gennari, *Tetrahedron*

Letters **1983**, *24*, 4405; h) W. S. Wadsworth, W. D. Emmons, *J. Am. Chem. Soc.* **1961**, *83*, 1733; i) G. Wittig, U. Schöllkopf, *Chem. Ber.* **1954**, *87*, 1318.

[127] a) H. Lindlar, *HCA* **1952**, *35*, 446; b) H. Lindlar, R. Dubuis, *Org. Synth.* **1966**, *46*, 89.

[128] a) J. A. Osborn, F. H. Jardine, J. F. Young, G. Wilkinson, *J. Chem. Soc., A* **1966**, 1711; b) R. R. Schrock, J. A. Osborn, *J. Am. Chem. Soc.* **1976**, *98*, 2143.

[129] a) R. Nishibayashi, T. Kurahashi, S. Matsubara, *Synlett* **2014**, *25*, 1287; b) E. STERN, *Journal of Catalysis* **1972**, *27*, 120; c) M. W. van Laren, C. J. Elsevier, *Angew. Chem. Int. Ed.* **1999**, *38*, 3715.

[130] G. F. Pregaglia, A. Andreetta, G. F. Ferrari, R. Ugo, *Journal of Organometallic Chemistry* **1971**, *30*, 387.

[131] C. Chen, Y. Huang, Z. Zhang, X.-Q. Dong, X. Zhang, *Chemical communications (Cambridge, England)* **2017**, *53*, 4612.

[132] S. C. Bart, E. Lobkovsky, P. J. Chirik, *J. Am. Chem. Soc.* **2004**, *126*, 13794.

[133] M. Sodeoka, M. Shibasaki, *J. Org. Chem.* **1985**, *50*, 1147.

[134] H. S. La Pierre, J. Arnold, F. D. Toste, *Angew. Chem. Int. Ed.* **2011**, *50*, 3900.

[135] T. L. Gianetti, N. C. Tomson, J. Arnold, R. G. Bergman, *J. Am. Chem. Soc.* **2011**, *133*, 14904.

[136] a) F. Pape, N. O. Thiel, J. F. Teichert, *Chemistry (Weinheim an der Bergstrasse, Germany)* **2015**, *21*, 15934; b) K. Semba, R. Kameyama, Y. Nakao, *Synlett* **2015**, *26*, 318; c) N. O. Thiel, J. F. Teichert, *Organic & biomolecular chemistry* **2016**, *14*, 10660; d) T. Wakamatsu, K. Nagao, H. Ohmiya, M. Sawamura, *Organometallics* **2016**, *35*, 1354.

[137] A. Brzozowska, L. M. Azofra, V. Zubar, I. Atodiresei, L. Cavallo, M. Rueping, O. El-Sepelgy, *ACS Catal.* **2018**, *8*, 4103.

[138] V. M. Chernyshev, A. V. Astakhov, I. E. Chikunov, R. V. Tyurin, D. B. Eremin, G. S. Ranny, V. N. Khrustalev, V. P. Ananikov, *ACS Catal.* **2019**, *9*, 2984.

[139] a) R. H. Crabtree, *Energy Environ. Sci.* **2008**, *1*, 134; b) P. Jessop, *Nature chemistry* **2009**, *1*, 350; c) L. Watson, O. Eisenstein, *J. Chem. Educ.* **2002**, *79*, 1269.

[140] a) K.-i. Fujita, Y. Tanaka, M. Kobayashi, R. Yamaguchi, *J. Am. Chem. Soc.* **2014**, *136*, 4829; b) R. Yamaguchi, C. Ikeda, Y. Takahashi, K.-i. Fujita, *J. Am. Chem. Soc.* **2009**, *131*, 8410.

[141] M. G. Manas, L. S. Sharninghausen, E. Lin, R. H. Crabtree, *Journal of Organometallic Chemistry* **2015**, *792*, 184.

[142] Á. Vivancos, M. Beller, M. Albrecht, *ACS Catal.* **2018**, *8*, 17.

[143] S. Wang, H. Huang, C. Bruneau, C. Fischmeister, *ChemSusChem* **2019**, *12*, 2350.

[144] R. Xu, S. Chakraborty, H. Yuan, W. D. Jones, *ACS Catal.* **2015**, *5*, 6350.

[145] a) A. Dubey, S. M. W. Rahaman, R. R. Fayzullin, J. R. Khusnutdinova, *ChemCatChem* **2019**, *11*, 3844; b) V. Papa, Y. Cao, A. Spannenberg, K. Junge, M. Beller, *Nat Catal* **2020**, *3*, 135; c) Y. Wang,

L. Zhu, Z. Shao, G. Li, Y. Lan, Q. Liu, *J. Am. Chem. Soc.* **2019**, *141*, 17337; d) Z. Wang, L. Chen, G. Mao, C. Wang, *Chinese Chemical Letters* **2020**.

[146] a) J. Buckingham, I. W. Southon, G. A. Cordell, *Dictionary of alkaloids*, **1989**; b) E. Fattorusso, O. Taglialatela-Scafati, *Modern alkaloids. Structure, isolation, synthesis and biology*, Wiley-VCH, Weinheim, Chichester, **2008**.

[147] H.-U. Blaser, H. Steiner, M. Studer, *ChemCatChem* **2009**, *1*, 210.

[148] D. Formenti, F. Ferretti, F. K. Scharnagl, M. Beller, *Chemical reviews* **2019**, *119*, 2611.

[149] M. Orlandi, D. Brenna, R. Harms, S. Jost, M. Benaglia, *Org. Process Res. Dev.* **2018**, *22*, 430.

[150] R. S. Downing, P. J. Kunkeler, H. van Bekkum, *Catalysis Today* **1997**, *37*, 121.

[151] A. Béchamp, *Annales de chimie et de physique* **1854**, *42*, 186.

[152] A. Corma, C. González-Arellano, M. Iglesias, F. Sánchez, *Applied Catalysis A: General* **2009**, *356*, 99.

[153] S. G. Harsy, *Tetrahedron* **1990**, *46*, 7403.

[154] Z. Yu, S. Liao, Y. Xu, B. Yang, D. Yu, *Journal of Molecular Catalysis A: Chemical* **1997**, *120*, 247.

[155] S. Xu, X. Xi, J. Shi, S. Cao, *Journal of Molecular Catalysis A: Chemical* **2000**, *160*, 287.

[156] E. G. Chepaikin, M. L. Khidekel', V. V. Ivanova, A. I. Zakhariev, D. M. Shopov, *Journal of Molecular Catalysis* **1980**, *10*, 115.

[157] a) A. A. Deshmukh, A. K. Prashar, A. K. Kinage, R. Kumar, R. Meijboom, *Ind. Eng. Chem. Res.* **2010**, *49*, 12180; b) A. Toti, P. Frediani, A. Salvini, L. Rosi, C. Giolli, *Journal of Organometallic Chemistry* **2005**, *690*, 3641.

[158] J. F. Knifton, *J. Org. Chem.* **1976**, *41*, 1200.

[159] R. M. Deshpande, A. N. Mahajan, M. M. Diwakar, P. S. Ozarde, R. V. Chaudhari, *J. Org. Chem.* **2004**, *69*, 4835.

[160] G. Wienhöfer, M. Baseda-Krüger, C. Ziebart, F. A. Westerhaus, W. Baumann, R. Jackstell, K. Junge, M. Beller, *Chemical communications (Cambridge, England)* **2013**, *49*, 9089.

[161] Borut ZUPANCIC, EP2981520A1, **2014**.

[162] J. Jampilek, *Molecules (Basel, Switzerland)* **2019**, *24*.

[163] a) Y. Tsuji, S. Kotachi, K. T. Huh, Y. Watanabe, *J. Org. Chem.* **1990**, *55*, 580; b) A. J.A. Watson, R. J. Wakeham, A. C. Maxwell, J. M.J. Williams, *Tetrahedron* **2014**, *70*, 3683; c) S. C. Ghosh, S. H. Hong, *Eur. J. Org. Chem.* **2010**, *2010*, 4266; d) Y. Tsuji, K.-T. Huh, Y. Yokoyama, Y. Watanabe, *J. Chem. Soc., Chem. Commun.* **1986**, 1575; e) T. Higuchi, R. Tagawa, A. Iimuro, S. Akiyama, H. Nagae, K. Mashima, *Chemistry (Weinheim an der Bergstrasse, Germany)* **2017**, *23*, 12795.

[164] a) K.-i. Fujita, K. Yamamoto, R. Yamaguchi, *Organic letters* **2002**, *4*, 2691; b) D. Pingen, D. Vogt, *Catal. Sci. Technol.* **2014**, *4*, 47; c) C.-F. Fu, Y.-H. Chang, Y.-H. Liu, S.-M. Peng, C. J. Elsevier, J.-T. Chen, S.-T. Liu, *Dalton transactions (Cambridge, England : 2003)* **2009**, 6991; d) T. Fukutake, K. Wada, G. C. Liu, S. Hosokawa, Q. Feng, *Catalysis Today* **2018**, *303*, 235; e) J.-Q. Li, P. G. Andersson, *Chemical communications (Cambridge, England)* **2013**, *49*, 6131; f) S. Huang, S.-P. Wu, Q. Zhou, H.-Z. Cui, X. Hong, Y.-J. Lin, X.-F. Hou, *Journal of Organometallic Chemistry* **2018**, *868*, 14.

[165] S. Gonell, M. Poyatos, J. A. Mata, E. Peris, *Organometallics* **2012**, *31*, 5606.

[166] T. Mori, C. Ishii, M. Kimura, *Org. Process Res. Dev.* **2019**, *23*, 1709.

[167] M. Peña-López, H. Neumann, M. Beller, *ChemCatChem* **2015**, *7*, 865.

[168] a) M. Vellakkaran, K. Singh, D. Banerjee, *ACS Catal.* **2017**, *7*, 8152; b) P. Yang, C. Zhang, W.-C. Gao, Y. Ma, X. Wang, L. Zhang, J. Yue, B. Tang, *Chemical communications (Cambridge, England)* **2019**, *55*, 7844.

[169] a) S. V. Samuelsen, C. Santilli, M. S. G. Ahlquist, R. Madsen, *Chemical science* **2019**, *10*, 1150; b) R. Fertig, T. Irrgang, F. Freitag, J. Zander, R. Kempe, *ACS Catal.* **2018**, *8*, 8525; c) L. Homberg, A. Roller, K. C. Hultzsch, *Organic letters* **2019**, *21*, 3142; d) V. G. Landge, A. Mondal, V. Kumar, A. Nandakumar, E. Balaraman, *Organic & biomolecular chemistry* **2018**, *16*, 8175; e) B. G. Reed-Berendt, L. C. Morrill, *J. Org. Chem.* **2019**, *84*, 3715; f) K. Das, A. Mondal, D. Pal, H. K. Srivastava, D. Srimani, *Organometallics* **2019**, *38*, 1815; g) M. Huang, Y. Li, Y. Li, J. Liu, S. Shu, Y. Liu, Z. Ke, *Chemical communications (Cambridge, England)* **2019**, *55*, 6213; h) U. K. Das, Y. Ben-David, Y. Diskin-Posner, D. Milstein, *Angew. Chem. Int. Ed.* **2018**, *57*, 2179; i) K. Das, A. Kumar, A. Jana, B. Maji, *Inorganica Chimica Acta* **2020**, *502*, 119358.

[170] a) G. Choi, S. H. Hong, *Angew. Chem. Int. Ed.* **2018**, *57*, 6166; b) G. Choi, S. H. Hong, *ACS Sustainable Chem. Eng.* **2019**, *7*, 716; c) M. Mastalir, E. Pittenauer, G. Allmaier, K. Kirchner, *J. Am. Chem. Soc.* **2017**, *139*, 8812; d) T. Kusakabe, *J. Syn. Org. Chem., Jpn.* **2018**, *76*, 622.

[171] a) K. Das, A. Mondal, D. Srimani, *J. Org. Chem.* **2018**, *83*, 9553; b) K. Das, A. Mondal, D. Pal, D. Srimani, *Organic letters* **2019**, *21*, 3223.

[172] G. E. Dobereiner, R. H. Crabtree, *Chemical reviews* **2010**, *110*, 681.

[173] a) X. Cui, Y. Deng, F. Shi, *ACS Catal.* **2013**, *3*, 808; b) X. Cui, Y. Zhang, F. Shi, Y. Deng, *Chemistry (Weinheim an der Bergstrasse, Germany)* **2011**, *17*, 2587; c) C. Feng, Y. Liu, S. Peng, Q. Shuai, G. Deng, C.-J. Li, *Organic letters* **2010**, *12*, 4888; d) C. Li, K.-F. Wan, F.-Y. Guo, Q.-H. Wu, M.-L. Yuan, R.-X. Li, H.-Y. Fu, X.-L. Zheng, H. Chen, *J. Org. Chem.* **2019**, *84*, 2158; e) Y. Liu, W. Chen, C. Feng, G. Deng, *Chemistry, an Asian journal* **2011**, *6*, 1142; f) B. Paul, S. Shee, K. Chakrabarti, S. Kundu, *ChemSusChem* **2017**, *10*, 2370; g) D.-W. Tan, H.-X. Li, D. J. Young, J.-P. Lang,

Tetrahedron **2016**, *72*, 4169; h) L. Wang, H. Neumann, M. Beller, *Angew. Chem. Int. Ed.* **2019**, *58*, 5417; i) S. Zhang, J. J. Ibrahim, Y. Yang, *Org. Chem. Front.* **2019**, *6*, 2726.

[174] B. G. Reed-Berendt, N. Mast, L. C. Morrill, *Eur. J. Org. Chem.* **2020**, *2020*, 1136.

[175] R. C. White, S. Ma, *Journal of Heterocyclic Chemistry* **1987**, *24*, 1203.

[176] a) Z. Shao, S. Fu, M. Wei, S. Zhou, Q. Liu, *Angewandte Chemie (International ed. in English)* **2016**, *55*, 14653; b) D. Spasyuk, D. G. Gusev, *Organometallics* **2012**, *31*, 5239.

[177] Q. Yao, *Organic letters* **2002**, *4*, 2197.

[178] J. Ortiz, A. Guijarro, M. Yus, *Eur. J. Org. Chem.* **1999**, *1999*, 3005.

[179] Y. Yang, J. Guo, H. Ng, Z. Chen, P. Teo, *Chemical communications (Cambridge, England)* **2014**, *50*, 2608.

[180] Q. Zhang, X. Kang, L. Long, L. Zhu, Y. Chai, *Synthesis* **2014**, *47*, 55.

[181] J. T. Colyer, N. G. Andersen, J. S. Tedrow, T. S. Soukup, M. M. Faul, *J. Org. Chem.* **2006**, *71*, 6859.

[182] M. Xiao, X. Yue, R. Xu, W. Tang, D. Xue, C. Li, M. Lei, J. Xiao, C. Wang, *Angew. Chem. Int. Ed.* **2019**, *58*, 10528.

[183] G. Borg, D. A. Cogan, J. A. Ellman, *Tetrahedron Letters* **1999**, *40*, 6709.

[184] V. Zubar, J. C. Borghs, M. Rueping, *Organic letters* **2020**, *22*, 3974.

[185] C. Belger, B. Plietker, *Chemical communications (Cambridge, England)* **2012**, *48*, 5419.

[186] Yuji NakayamaOsamu Ogata, US10072033B2.

[187] L. Ilies, T. Yoshida, E. Nakamura, *J. Am. Chem. Soc.* **2012**, *134*, 16951.

[188] D. C. Fabry, M. A. Ronge, M. Rueping, *Chemistry (Weinheim an der Bergstrasse, Germany)* **2015**, *21*, 5350.

[189] K. Li, R. Khan, X. Zhang, Y. Gao, Y. Zhou, H. Tan, J. Chen, B. Fan, *Chemical communications (Cambridge, England)* **2019**, *55*, 5663.

[190] M. Das, D. F. O'Shea, *Organic letters* **2016**, *18*, 336.

[191] X. Han, J. Hu, C. Chen, Y. Yuan, Z. Shi, *Chemical communications (Cambridge, England)* **2019**, *55*, 6922.

[192] S. Hancker, H. Neumann, M. Beller, *Eur. J. Org. Chem.* **2018**, *2018*, 5253.

[193] R. Iwasaki, E. Tanaka, T. Ichihashi, Y. Idemoto, K. Endo, *The Journal of organic chemistry* **2018**, *83*, 13574.

[194] L. P. Szajek, J. R. Shapley, *Organometallics* **1994**, *13*, 1395.

[195] T. Nakagiri, M. Murai, K. Takai, *Organic letters* **2015**, *17*, 3346.

[196] Y. S. Wagh, N. Asao, *The Journal of organic chemistry* **2015**, *80*, 847.

[197] J. C. Borghs, V. Zubar, L. M. Azofra, J. Sklyaruk, M. Rueping, *Organic letters* **2020**, *22*, 4222.

[198] A. Kulkarni, W. Zhou, B. Török, *Organic letters* **2011**, *13*, 5124.
[199] F. Chen, A.-E. Surkus, L. He, M.-M. Pohl, J. Radnik, C. Topf, K. Junge, M. Beller, *J. Am. Chem. Soc.* **2015**, *137*, 11718.
[200] G. Chessari, I. M. Buck, J. E. H. Day, P. J. Day, A. Iqbal, C. N. Johnson, E. J. Lewis, V. Martins, D. Miller, M. Reader et al., *Journal of medicinal chemistry* **2015**, *58*, 6574.
[201] H.-Y. Lee, L.-T. Wang, Y.-H. Li, S.-L. Pan, Y.-L. Chen, C.-M. Teng, J.-P. Liou, *Organic & biomolecular chemistry* **2014**, *12*, 8966.
[202] F. Ding, Y. Zhang, R. Zhao, Y. Jiang, R. L.-Y. Bao, K. Lin, L. Shi, *Chemical communications (Cambridge, England)* **2017**, *53*, 9262.
[203] E. L. Piatnitski Chekler, R. Unwalla, T. A. Khan, R. S. Tangirala, M. Johnson, M. St Andre, J. T. Anderson, T. Kenney, S. Chiparri, C. McNally et al., *Journal of medicinal chemistry* **2014**, *57*, 2462.
[204] K. Sato, H. Sugimoto, K. Rikimaru, H. Imoto, M. Kamaura, N. Negoro, Y. Tsujihata, H. Miyashita, T. Odani, T. Murata, *Bioorganic & medicinal chemistry* **2014**, *22*, 1649.
[205] C. Li, J. Chen, G. Fu, D. Liu, Y. Liu, W. Zhang, *Tetrahedron* **2013**, *69*, 6839.
[206] L. Zhang, R. Qiu, X. Xue, Y. Pan, C. Xu, H. Li, L. Xu, *Adv. Synth. Catal.* **2015**, *357*, 3529.
[207] N. Zumbrägel, P. Machui, J. Nonnhoff, H. Gröger, *The Journal of organic chemistry* **2019**, *84*, 1440.
[208] V. Kanchupalli, D. Joseph, S. Katukojvala, *Organic letters* **2015**, *17*, 5878.
[209] J. Zhang, S. Chen, F. Chen, W. Xu, G.-J. Deng, H. Gong, *Adv. Synth. Catal.* **2017**, *359*, 2358.
[210] K. Chakrabarti, M. Maji, S. Kundu, *Green Chem.* **2019**, *21*, 1999.
[211] Y. Mao, Y. Liu, Y. Hu, L. Wang, S. Zhang, W. Wang, *ACS Catal.* **2018**, *8*, 3016.
[212] J. D. Al-Shawabkeh, A. H. Al-Nadaf, L. A. Dahabiyeh, M. O. Taha, *Med Chem Res* **2014**, *23*, 127.
[213] C. E. Katz, J. Aubé, *J. Am. Chem. Soc.* **2003**, *125*, 13948.
[214] E. Baslé, M. Jean, N. Gouault, J. Renault, P. Uriac, *Tetrahedron Letters* **2007**, *48*, 8138.
[215] H. R. Suryavanshi, M. M. Rathore, *Org. Commun.* **2017**, *10*, 228.
[216] S. Sharma, Y. Yamini, P. Das, *New J. Chem.* **2019**, *43*, 1764.
[217] J. Gao, S. Bhunia, K. Wang, L. Gan, S. Xia, D. Ma, *Organic letters* **2017**, *19*, 2809.
[218] P. G. Alsabeh, R. J. Lundgren, R. McDonald, C. C. C. Johansson Seechurn, T. J. Colacot, M. Stradiotto, *Chemistry (Weinheim an der Bergstrasse, Germany)* **2013**, *19*, 2131.
[219] J. Yu, P. Zhang, J. Wu, Z. Shang, *Tetrahedron Letters* **2013**, *54*, 3167.
[220] S. Lee, M. Jørgensen, J. F. Hartwig, *Organic letters* **2001**, *3*, 2729.
[221] P. Ryabchuk, K. Junge, M. Beller, *Synthesis* **2018**, *50*, 4369.
[222] K. D. Janda, J. A. Ashley, T. M. Jones, D. A. McLeod, D. M. Schloeder, M. I. Weinhouse, R. A. Lerner, R. A. Gibbs, P. A. Benkovic, *J. Am. Chem. Soc.* **1991**, *113*, 291.

[223] A. Ojeda-Porras, A. Hernández-Santana, D. Gamba-Sánchez, *Green Chem.* **2015**, *17*, 3157.
[224] H.-J. Xu, Y.-F. Liang, Z.-Y. Cai, H.-X. Qi, C.-Y. Yang, Y.-S. Feng, *The Journal of organic chemistry* **2011**, *76*, 2296.
[225] L. Lu, J. Ma, P. Qu, F. Li, *Organic letters* **2015**, *17*, 2350.
[226] X.-G. Song, Y.-Y. Ren, S.-F. Zhu, Q.-L. Zhou, *Adv. Synth. Catal.* **2016**, *358*, 2366.
[227] A. Ohtaka, M. Kozono, K. Takahashi, G. Hamasaka, Y. Uozumi, T. Shinagawa, O. Shimomura, R. Nomura, *Chem. Lett.* **2016**, *45*, 758.
[228] M. J. Islam, K. Matsuo, H. M. Menezes, M. Takahashi, H. Nakagawa, A. Kakugo, K. Sada, N. Tamaoki, *Organic & biomolecular chemistry* **2018**, *17*, 53.
[229] K.-i. Fujita, Y. Takahashi, M. Owaki, K. Yamamoto, R. Yamaguchi, *Organic letters* **2004**, *6*, 2785.
[230] Pascale Gaillard, Vincent Pomel, Isabelle Jeanclaude-Etter, Jérôme DORBAIS, Jasna Klicic, Cyril Montagne, WO2008101979A1.
[231] J. C. A. Flanagan, L. M. Dornan, M. G. McLaughlin, N. G. McCreanor, M. J. Cook, M. J. Muldoon, *Green Chem.* **2012**, *14*, 1281.
[232] Y. Konda-Yamada, C. Okada, K. Yoshida, Y. Umeda, S. Arima, N. Sato, T. Kai, H. Takayanagi, Y. Harigaya, *Tetrahedron* **2002**, *58*, 7851.
[233] P. Ye, Y. Shao, X. Ye, F. Zhang, R. Li, J. Sun, B. Xu, J. Chen, *Organic letters* **2020**, *22*, 1306.
[234] Y.-N. Duan, X. Du, Z. Cui, Y. Zeng, Y. Liu, T. Yang, J. Wen, X. Zhang, *J. Am. Chem. Soc.* **2019**, *141*, 20424.
[235] Y. Wang, B. Dong, Z. Wang, X. Cong, X. Bi, *Organic letters* **2019**, *21*, 3631.
[236] K. Kolmakov, E. Hebisch, T. Wolfram, L. A. Nordwig, C. A. Wurm, H. Ta, V. Westphal, V. N. Belov, S. W. Hell, *Chemistry (Weinheim an der Bergstrasse, Germany)* **2015**, *21*, 13344.
[237] S. Stockerl, T. Danelzik, D. G. Piekarski, O. García Mancheño, *Organic letters* **2019**, *21*, 4535.
[238] T. V. Nykaza, J. C. Cooper, G. Li, N. Mahieu, A. Ramirez, M. R. Luzung, A. T. Radosevich, *J. Am. Chem. Soc.* **2018**, *140*, 15200.
[239] P. Weber, T. Scherpf, I. Rodstein, D. Lichte, L. T. Scharf, L. J. Gooßen, V. H. Gessner, *Angew. Chem. Int. Ed.* **2019**, *58*, 3203.
[240] W. Zhou, M. Fan, J. Yin, Y. Jiang, D. Ma, *J. Am. Chem. Soc.* **2015**, *137*, 11942.